11th International Conference on Building Defects (Building Defects 2019)

IOP Conference Series: Materials Science and Engineering
Volume 728

Ceske Budejovice, Czech Republic
28 - 29 November 2019

ISBN: 978-1-7138-2587-6
ISSN: 1757-8981

Printed from e-media with permission by:

Curran Associates, Inc.
57 Morehouse Lane
Red Hook, NY 12571

Some format issues inherent in the e-media version may also appear in this print version.

Additional copies of this publication are available from:

Curran Associates, Inc.
57 Morehouse Lane
Red Hook, NY 12571 USA
Phone: 845-758-0400
Fax: 845-758-2633
Email: curran@proceedings.com
Web: www.proceedings.com

TABLE OF CONTENTS

DYNAMICAL MODEL FOR CALCULATION TRANSVERSE FORCES BETWEEN OVERHEAD BRIDGE CRANE AND CRANE RUNWAY 1
Josef Musilek

ABSORBENT PROPERTIES OF POLYSTYRENE WITH CLOSED OUTER STRUCTURE IN CASE OF FAILURE OF THE EXTERNAL SURFACE OF THE BOARD 7
Ondrej Pilný, Lubor Kalousek

SURVEY OF THE DESIGNS OF WOODEN ARCHITECTURAL MONUMENTS 14
E N Pokrovskaya

DIAGNOSTICS OF FLAT ROOFS WITH FLEXIBLE SHEETS FOR WATERPROOFING 20
Stanislav Šutliak, Jan Plachý

SECONDARY RAW MATERIALS IN CONSTRUCTION 27
Radim Smolka, Jindrich Sobotka

SURFACE TEMPERATURE OF FLAT ROOFS WITH WATERPROOFING POLYMER MEMBRANES 33
Jan Plachý, Jana Vysoká

HEATING AND DRYING OF AUTOCLAVED AERATED CONCRETE BLOCKS FOR DEFECTS OF BUILDINGS BY APPLICATION OF MICROWAVE RADIATION 37
Jindrich Sobotka, Radim Smolka

DEGRADATION OF PVC WATERPROOFING MEMBRANE CONTAINING PLASTICIZERS 42
Marek Novotný

MID-RISE ATRIUM RESIDENTIAL HOUSES - DOMAIN OF COMFORT 48
Irina Maltseva, Natalia Kaganovich, Konstantin Tkachuk

SMART GREEN BRIDGE - WILDLIFE CROSSING BRIDGES OF NEW GENERATION 54
Jaroslav Žák, Michal Kraus, Petra Machová, Jan Plachý

REHABILITATION AND REINFORCEMENT OF FERROCONCRETE WALLS WITH REINFORCED PILASTERS 60
Michal Novotný, Jan Všetecka

LEVEL OF TOTAL VOLATILE ORGANIC COMPOUNDS (TVOC) IN THE CONTEXT OF INDOOR AIR QUALITY (IAQ) IN OFFICE BUILDINGS 68
Michal Kraus, Ingrid Juhásová Šenitková

USE OF INFORMATION MODEL IN CREATING THE INVESTMENT-CONSTRUCTION PROJECT 75
V M Cherkina, N S Shushunova

APPLICATION OF MODELING PROCESSES AND 3D PRINT ON CASTING MOLDS FOR CONCRETE FURNITURE 82
Jirí Šál, Martin Dedic

METHODOLOGICAL FEATURES OF THE MANAGEMENT OF MODERN LABOUR PROTECTION IN THE CONSTRUCTION 87
E B Sugak

DISORDERS OF THE GREEN ROOF OF THE POOL LUCENICE 95
Martin Dedic, Aleš Kankovský

EXPERIMENTAL MEASUREMENTS OF WOODEN SOLID LOG WALL 101
Petr Kacálek, Tomáš Petrícek

REDUCTION OF LIGHT TRANSMISSION BY GLAZING WITH ATMOSPHERIC POLLUTANTS 106
Petra Machová, František Vajkay

Author Index

IOP Conf. Series: Materials Science and Engineering **728** (2020) 011001 doi:10.1088/1757-899X/728/1/011001

Building Defects 2019

Ingrid Juhásová Šenitková [1]

[1] Institute of Technology and Business in České Budějovice, Department of Civil Engineering, Okružní 517/10, 370 01 České Budějovice, Czech Republic

Infobuildingdefects@gmail.com

Preface

We are pleased to welcome you to the conference proceedings of the 11th Anniversary International Scientific Conference on Building Defects (Building Defects 2019) held on 28th – 29th November 2019 at the Institute of Technology and Business (ITB) in České Budějovice, Czech Republic. Conference is organized on behalf of His Magnificence prof. Marek Vochozka, MSc., MBA., Ph.D., rector of the Institute of Technology and Business in České Budějovice and under the honorary patronage of the Czech Chamber of Certified Engineers and Technicians (CKAIT).

The current conference is the eleventh continuous of Building Defects conferences, which solves especially the causes of defects during the building processes and afterwards defects of the building themselves as well as the defect of building structures. The conference is the premiere chance for the presentation of new advances and research results in the fields of the civil engineering. The main aim of the conference is to the present the scientific research results and its practical solutions in the field of civil engineering, taking into account the principles of sustainable development as well as to present contemporary civil engineering issues and its future development.

It is our pleasure to announce the participation of expert speakers from various universities and research institutes in this two-day event. A lot of thanks are given to all authors of submitted papers for their participation. They contributed a great deal of effort and creativity to produce this work. We are happy they choose the 11th Anniversary International Scientific Conference on Building Defects to present to their excellent work. Thanks also to all editors and reviewers, who donated substantial time from their busy schedules to carefully read and evaluate the submitted scientific papers.

The conference proceedings contain 20 papers. Each contributed paper was rigorously reviewed by two anonymous reviewers who were drawn mainly from the International Scientific Committee. At last but not at least, we would like to appreciate to all supporting organizations for their support and encouragement and got making the event a success. In addition, I would like to thank also to Organizing Committee for their great preparing work. We are looking forward to the next 12th International Conference Building Defects 2020 that will be held next year on November at the same location. We hope that it will be interesting and enjoying at least as all of its eleventh predecessors.

prof. Ingrid Juhásová Šenitková, MSc., CSc.
Conference Guarantee
ITB in České Budějovice

IOP Conf. Series: Materials Science and Engineering **728** (2020) 011002 doi:10.1088/1757-899X/728/1/011002

Peer review statement

All papers published in this volume of *IOP Conference Series: Materials Science and Engineering* have been peer reviewed through processes administered by the proceedings Editors. Reviews were conducted by expert referees to the professional and scientific standards expected of a proceedings journal published by IOP Publishing.

IOP Conf. Series: Materials Science and Engineering **728** (2020) 012001 doi:10.1088/1757-899X/728/1/012001

Dynamical Model for Calculation Transverse Forces between Overhead Bridge Crane and Crane Runway

Josef Musilek[1]

[1] Institute of Technology and Business in České Budějovice, Department of Civil Engineering, Okružní 517/10, 370 01 České Budějovice, Czech Republic

musilek@mail.vstecb.cz

Abstract. Horizontal forces between the crane and the crane runway girder occur during a motion of an overhead travelling crane on the crane runway. There are several methods for calculation of these forces. The most significant load caused by motion of crane is the skewing and the acceleration of the crane and of the crab. This paper deals with the acceleration of the crane and with the skewing of the crane. In present the skewing of crane is defined as the motion of the overhead bridge crane with the constant velocity but with the angle relative to the crane runway. The transverse forces caused by acceleration of the crane are calculated separately. But the skewing of the crane can occur also during the accelerating of the crane. Present standards don't take this situation into account. A dynamical model presented in this paper describes motion of the overhead double bridge crane during its acceleration. The basic assumption of this model is that there the contact between the rim of the firs crane wheel and the rail of the crane runway.

1. Introduction

The transverse horizontal forces between the overhead bridge crane and the rail of crane runway girder occur during a motion of the crane. These forces are caused by various causes. They can be caused by accelerating or decelerating of crane, acceleration or deceleration of crab and the by the skewing of the crane which results into significant transverse forces between the crane wheel and the rail of the crane runway. There are some procedures and computational models which enable to calculate these horizontal transverse forces. Some of them were presented and compared in [1]. The comparison of different models shows the big differences between models and results. These big differences are caused by different interpretations of what is skewing of crane. N. A. Lobov [2-4] created dynamical model describing motion of the overhead bridge crane. This model enables to determinate the horizontal transverse forces. The model considers the crane as the rigid body with the load of crab rigidly connected to the crane. It means the swinging of the load on the crab rope was neglected.

The dynamical model created and presented in [5] considers some factors that Lobov's model did not consider. This includes the deformation of crane bridges (so-called S-deformation), the swinging of the crab load in the direction of the crane and the inclusion of crane drive control. Model is designed for a double bridge crane. The model is made up of rigid bodies that are connected to each other by equivalent springs, which represent the system's deformations. The motion equations of the

IOP Conf. Series: Materials Science and Engineering **728** (2020) 012001 doi:10.1088/1757-899X/728/1/012001

model are assembled using Lagrange equations II. kinds. The basic assumption of this model is that there is no contact between the rim of the crane wheel and the rail of the crane runway.

2. Dynamical Model of the overhead double bridge crane

The dynamical model presented in this paper is based on the model presented in [5]. The dynamical model in this paper assumes the contact between the rim of the crane wheel and the rail of the crane runway. The factors like deformation of crane bridges (so-called S-deformation), the swinging of the crab load in the direction of the crane and the inclusion of crane drive control are still taken into account.

Figure 1shows the created dynamical model which takes the contact between the rim of the crane wheel and the rail of the crane runway.

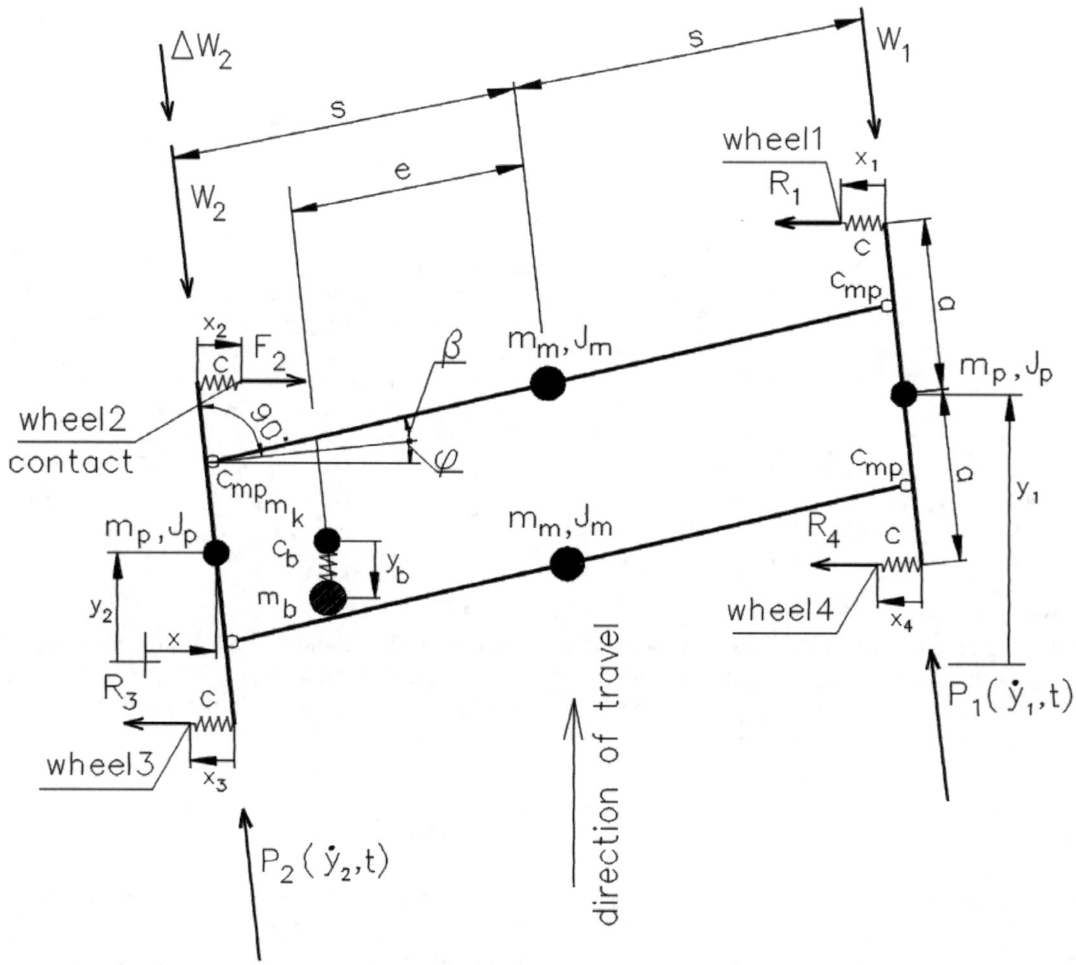

Figure 1. Dynamical model of the overhead bridge crane with contact of the wheel rim.

The symbols used in Figure 1 are explained in [5]. But compared to the model without the contact of the wheel rim and with the contact presented on Figure 1in this paper, some new symbols appeared and some symbols are different.

The situation for wheel 2 has changed, where force F_2 acts instead of force R_2. In this case, the force F_2 is the force that loads the crane runway girder and the crane structure. The situation of the forces acting on the wheel at location 2 is illustrated on the Figure 2.

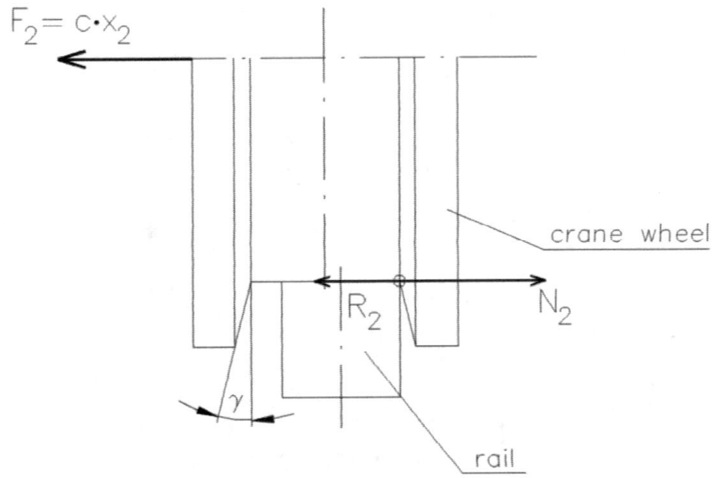

Figure 2. Forces acting on the crane wheel 2.

Legend to Figure 2:

- N_2 the contact force between the guide means (the rims or the guide rollers) and the rail side. This contact force is important in terms of wear of the rims of the wheels or of dimensioning of the guide rollers
- R_2 slip force between wheel and rail
- F_2 the force acting by the crane on the wheel

It's a fact, that:

$$N_2 = R_2 + F_2 \tag{1}$$

As a result, the loading of the guide means is higher than the loading of the crane rail and the crane structure when the crane is inclined on the crane runway.

The force ΔW_2 occurs in this computational model compared to the model without the contact of the rim. This force is caused by the friction of the guide means along the side of the track. In the case where the guiding means are realized like rims, the force ΔW_2 shall be determined from the following formula which is given in [2]:

$$\Delta W_2 = N_2 \cdot \frac{f \cdot |\varphi + \beta_2|}{tg\gamma} \tag{2}$$

where:

- N_2 the contact force between the wheel rim and the rail side
- R_2 coefficient of friction between the wheel rim and the rail side
- φ crane rotation according to calculation model

- β_2 the angle between the wheel in the end carriage measured between the end carriage and the crane wheel
- γ angle of deflection of the trims from the rail side

In [6] there is presented another formula that can be used to calculate the friction force ΔW_2 in case the guiding elements are the rims:

$$\Delta W_2 = N_2 \cdot f \cdot \frac{\frac{z^2}{R_k^2} + (\varphi + \beta_2)^2 \cdot (\tan(\frac{\pi}{2} - \gamma))^2}{\sqrt{\frac{z^2}{R_k^2} \cdot (\sin(\frac{\pi}{2} - \gamma))^2 + (\varphi + \beta_2)^2 \cdot (\tan(\frac{\pi}{2} - \gamma))^2}} \quad (3)$$

The meaning of the symbols in formula (3) is the same like in formula (2). But in formula (3) two more symbols occurred:

- R_k radius of crane wheel
- z depth of contact of the rim on the side of the rail – see Figure 3

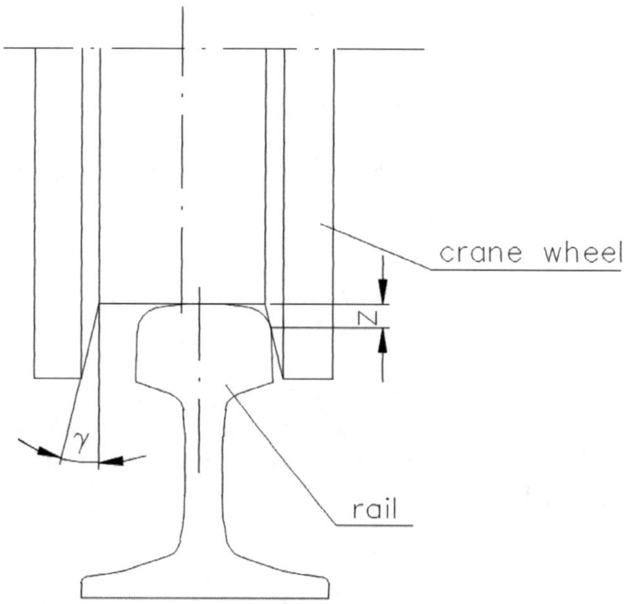

Figure 3. Depth of contact of the rim on the side of the rail.

Comparison of formulas (2) and (3) was made. The result is shown in the graph in Figure 4 (input values: $f = 0.3$, $\gamma = 0.175$ *rad*, $R_k = 0.185$ *m*, $N_2 = 1$ *N*, the value $(\varphi + \beta 2)$ is like whole variable). If $z = 0$ is entered in formula (3) (which would be true, for example, for a rail made of a square cross section without chamfering), relation (3) coincides with relation (2). The relationship (3) is used in the model.

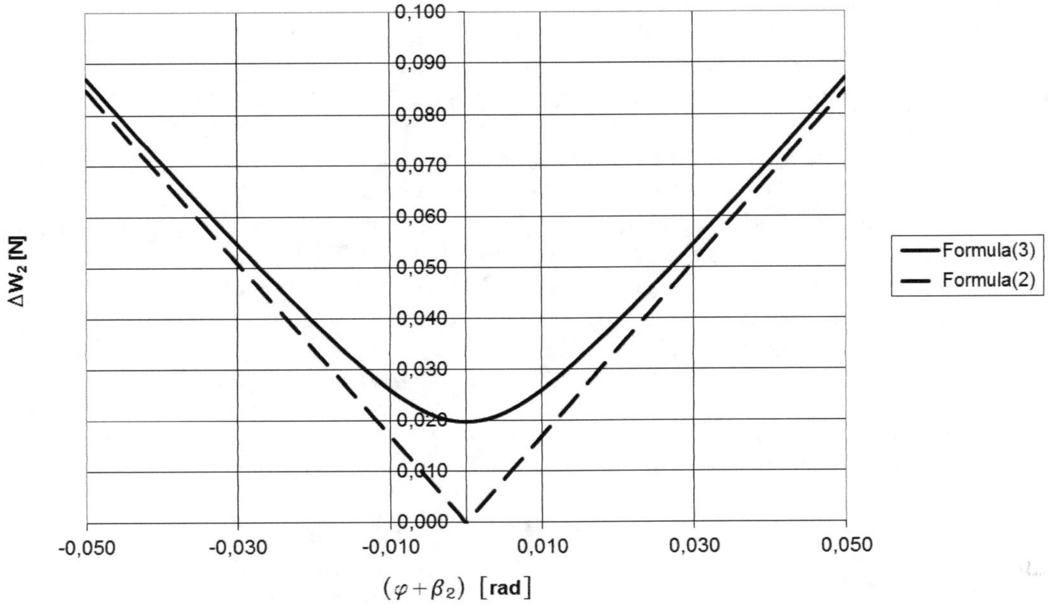

Figure 4. Comparison of formulas (2) and (3).

3. Motion equations of the dynamical model

The creation of motion equations is explained in [5]. The motion equations form model presented in this paper are the same like for the model presented in [5] but due to the contact of the rim at the wheel 2, the model lost one degree of freedom. The model now has 8 degrees of freedom. The variable *x2* is no longer unknown and can be expressed using the following condition:

$$x_2 = a \cdot \varphi - x - \delta \tag{4}$$

Where:

- δ is the half clearance between the rim and rails
- a, φ, x variables according to Figure 1

The formulas (3) and (4) complete the system of motions equations for dynamical model with the contact between the rim of the crane wheel and the rail of the crane runway.

4. Results and discussions

The dynamical model presented here is the extension of the dynamical model created by the author in [5]. The model presented in this paper assumed the contact between the rim of the crane wheel and the rail of the crane runway. The combination of these two models with the contact of the rim and without the contact of the rim enables to simulate the behavior of the crane moving on the crane runway. These two models enable to investigate not only the movement of the crane in general, but also the magnitude of the transverse forces acting between the wheels and the rail of the crane runway.

5. Conclusions

The extension of dynamical model created by author describing the motion of the overhead bridge crane on the crane runway was created. The construction of the extended model was explained and the extension of motion equations was shown. The main meaning of model extension is in considering contact between the the rim of the crane wheel and the rail of the crane runway.

References

[1] J. Musilek, "Horizontal Forces on Crane Runway Caused by Skewing of the Crane" *IOP Conference Series: Materials Science and Engineering*, vol. 471, issue 5, 052001, 2019.
[2] N. A. Lobov, "Loads of an overhead travelling crane caused by transverse and rotatory motions of the bridge girder" *Vestnik Mashinostroeniya*, vol. 64, pp. 22–26, 1984.
[3] N. A. Lobov, "Loads on an overhead travelling crane when it moves with a constant skew setting of the girder" *Vestnik Mashinostroeniya*, vol. 66, pp. 13–17, 1986.
[4] N. A. Lobov, A. V. Masyagin and A. V. Dulev, "On lengthening the life of bridge crane track wheels," *Vestnik Mashinostroeniya*, vol. 69, pp. 30–34, 1989.
[5] J. Musilek, "Dynamical Model for Determination of Horizontal Forces on Crane Runway during Motion of the Crane," *IOP Conference Series: Materials Science and Engineering*, vol. 603, issue 5, 052076, 2019.
[6] V. Švejnoch, *Theory of rail vehicles*. ČVUT Prague, 1991.

Absorbent Properties of Polystyrene with Closed Outer Structure in Case of Failure of the External Surface of the Board

Ondřej Pilný[1], Lubor Kalousek[1]

[1] Institute of Building Structures, Brno University of Technology, Faculty of Civil Engineering, Veveří 331/95, 602 00 Brno, Czech Republic

pilny.o@fce.vutbr.cz

Abstract. Thermal insulation is an integral part of almost all modern buildings. An important aspect in this process is the elimination of thermal bridges and insulation of all structures which are exposed to adverse environmental conditions. We are increasingly encountering thermal insulation solutions for elements that have been neglected in the past. One of these constructions is the substructure, which means not only the foundations, but underground floors and plinth. These constructions must be insulated with a material which not only has suitable thermal-insulating properties, but also has to withstand the effects of moisture in the long term. Nowadays, expanded polystyrene with a closed outer structure has become one of the most commonly used thermal insulators for these purposes. However, it is clear that, when built-in, the outer structure of the envelope is often damaged. This is often neglected phenomenon, which has a negative effect in the amount of diffusion flow and moisture propagation in the material. This paper describes a laboratory experiment, in which the outer structure of the EPS samples was broken in various ways. The samples were then fully immersed in water, changes in mass and volume absorbency over time were recorded and results were reciprocally compared.

1. Introduction

Thermal insulation as a protection of structures against heat leakage and its proper design is today, besides ensuring a static balance of structures and interior protection against the influence of moisture, one of the main duties of each designer and contractor. As a result of the stricter thermal standards in Europe [1], we are increasingly encountering thermal insulation of structures that have been neglected in the past due to technical difficulty when performing insulation or because of their apparent lack of importance for the overall heat loss. For example, these are not only penetrations through constructions in the form of installations, but mainly constructions of substructure, foundation structures, walls of underground floors, or, last but not least, a plinth.

Modern computational methods for the simulation of heat flow and conduction using the MESH and CFD methods, together with thermal imaging techniques, however shows us the influence of these thermal bridges on structures that seemed immaterial in past [2]. Construction nowadays takes the protection of these structures almost as a matter of course. However, the problem often arises in choosing a suitable thermal insulator.

In this environment, i.e. the substructure, the materials must meet several properties. In addition to a suitable thermal conductivity coefficient λ [$Wm^{-1}K^{-1}$], the materials must resist mechanical damage, be dimensionally stable and be effectively resistant to the effects of moisture over time. The longer the thermal insulator is exposed to moisture, the more it deteriorates its heat-insulating property, which depends on the thermal conductivity coefficient λ. It is therefore necessary to choose materials that are as non-absorbent as possible. One of these materials is, for example, currently commonly used expanded polystyrene (abbreviation in civil engineering EPS) with a closed outer structure, i.e. a thermal insulator, which obtains its resistance against humidity by outer structure of the envelope of individual boards and which is designed for structures in direct contact with moisture and high loads. It is however clear that when installed in a conventional building process, said board envelope mechanically breaks down either by mistake or for the purpose of cutting and subsequently gluing + anchoring to the structures in question.

The main object of this paper is to describe a laboratory experiment in which samples of expanded polystyrene with closed outer structure of the board were subjected to various surface treatments commonly found on the building site, i.e. cutting wire (hot-wire cutter) and polystyrene saw. The individual surface treatments were then compared with the conventional unmodified board structure.

2. Used method, materials, boundary conditions and samples measurements

2.1. Used Method

Within the complexity of measurement, standard methods were used to compute the volume absorption capacity for thermal insulation products for use in the construction industry according to ČSN EN 12087 [3] and weight absorption according to the standard ČSN 72 2603 [4]. These methods are used to determine the long-term absorption of materials during immersion. The method of immersion was considered as complete with upper surface of sample about 50 mm below water level for 28 days.

2.2. Materials and Samples

A total of 12 samples were prepared for the laboratory experiment in the selected method with dimensions of the sides of the base about 200 x 200 mm. The expanded polystyrene with a closed outer structure as a material was used due to very low water absorption. The material is intended for environments with direct contact with moisture. Since it is not normally possible to get samples of these exact dimensions needed for the method, the samples were made from larger boards. The samples were prepared so that the dimensions of the base correspond with the standards prescribed. This gave rise to cubes of the prescribed dimensions, which were suitably adjusted from 4 sides to achieve the required criteria and the two opposite sides (see Figure 1) were left as the sides on which the effect of surface treatments was measured.

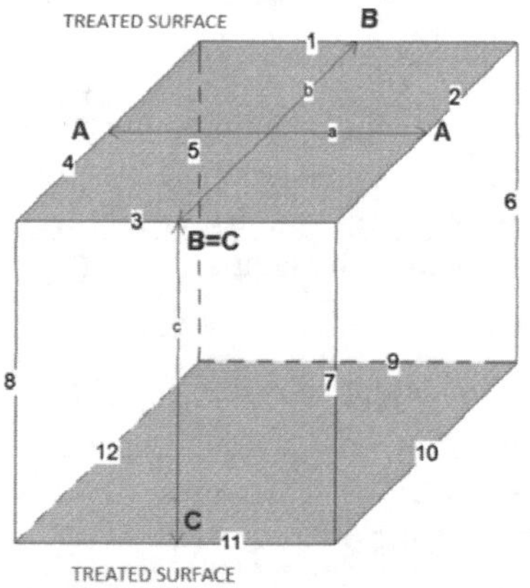

Figure 1. Sample Diagram [5].

Figure 2. Samples before Asphalt Dispersion Application [5].

2.3. Boundary Conditions

The boundary conditions on the laboratory corresponded to the prescribed values for the method during the course of the experiment. A water level of 50 ± 2 mm was maintained after immersion. A water temperature of 23 ± 5°C was maintained in the laboratory environment. Due to the creation of unsuitable boundary conditions on non-subject cutting surfaces (see Figure 2) in the creation of prescribed dimensions of individual samples, the effect of the mentioned shape and surface treatments was eliminated by coating the surfaces with dispersed asphalt mixture with modified synthetic rubber with thixotropic properties (hereinafter rubber-asphalt), which is intended for polystyrene. This coating (see Figure 3) was applied in two layers with a total thickness of approx. 2 ~ 2.5 mm. Individual samples were then allowed to dry for 3 days to eliminate residual moisture in the rubber-asphalt coating. After thorough drying, the samples were divided according to the test surface:

1. **Test Samples A** – 4 samples were treated in the subject sides with a cutting wire. Samples were numbered 1 ~ 4.
2. **Test Samples B** – 4 samples were treated in the subject sides with a polystyrene saw. Samples were numbered 5 ~ 8.
3. **Test Samples C** – 4 samples were left with the original board structure without any surface treatment on the examined subject sides. Samples were numbered 9 ~ 12.

Figure 3. Samples after Asphalt Dispersion Application [5].

The intrinsic effect of the absorbency of rubber-asphalt has not been investigated in detail. For the experiment and results, the absorption of expanded polystyrene samples over time and the subsequent comparison of individual samples with different surface treatment methods A – C were decisive. The water absorption of asphalt waterproofing is not commonly stated by the manufacturer and can be found in other scientific publications [6].

2.4. Samples Measurements

The samples were measured using a digital calliper with an accuracy of 0.03 – 0.04 mm and weighted with a digital scales of 0.1 g accuracy before being immersed in water. Subsequently, the samples were suitably immersed in a container of water at the prescribed temperature, suitably burdened and placed in a laboratory environment with constant conditions prescribed by the standard. The samples were withdrawn every 24 hours, suitably dried only from surface moisture, re-measured in places designated as A-A, B-B and C-C and weighted to determine the effect not only on water absorption, but also on volume stability.

3. Results

3.1 Used Equations

The following formulas for volume (1) and mass absorbency (2) were used to calculate and compare the absorbent properties. These formulas were derived from formulas given in the present standards:

$$n_v = \frac{m_n - m_s}{V_{28}} \times \frac{100}{\rho_k} \ [\%] \quad (1)$$

where: n_V long-term volume water absorption [%]
V_{28} volume of test sample after 28 days, $V_{28}=V_0$ [m^3]
m_s initial sample weight in time 0 [kg]
m_n weight of the sample when exposed to moisture after 28 days of immersion [kg]
ρ_k density of water [$kg.m^{-3}$]

$$n_h = \frac{m_n - m_s}{m_s} \times 100 \ [\%] \quad (2)$$

where: n_h long-term weight water absorption [%]
m_n weight of the sample when exposed to moisture after 28 days of immersion [kg]
m_s initial sample weight in time 0 [kg]

Individual equations were chosen with regard to the most accurate evaluation of the investigated properties. Formula (1) was corrected for the difference in the initial (V_0) and final (V_{28}) volume of the resulting values on the percentage comparison between these values. These values are based on the measurements of dimensions A-A = 0.52%, B-B = 0.34% a C-C = 1.08 %, where individual values were calculated from the minimum and maximum average values.

3.2 Resulting Graphs

The resulting values were summarized in graphs showing the course of water absorption over time.

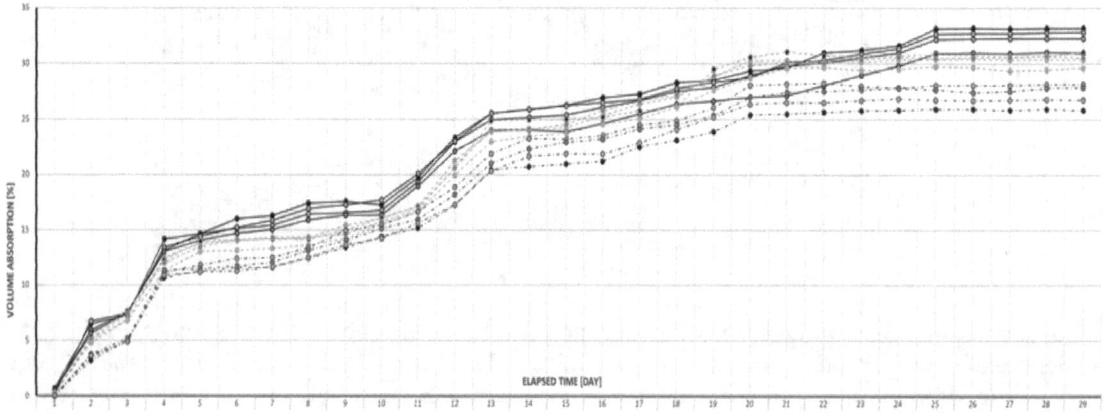

Figure 4. Graph of Volume Absorption of Polystyrene with Different Surface Treatment.

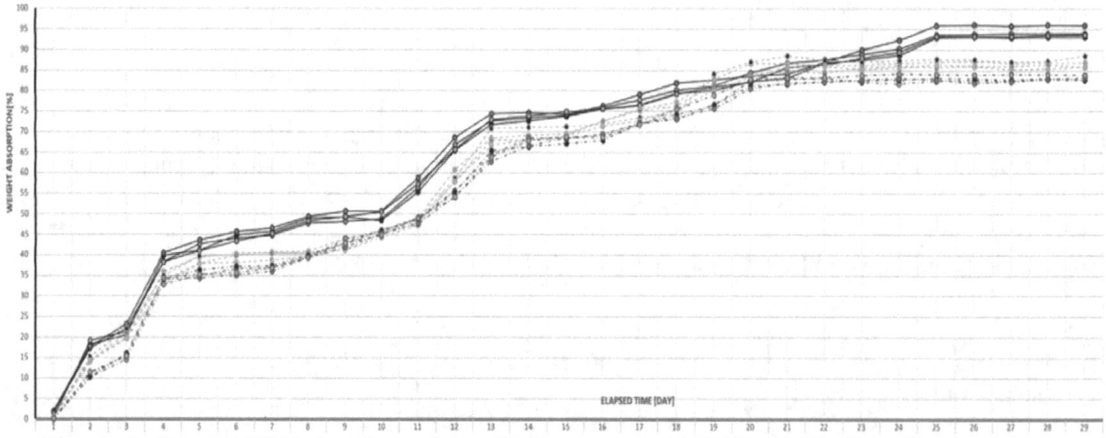

Figure 5. Graph of Weight Absorption of Polystyrene with Different Surface Treatment.

IOP Conf. Series: Materials Science and Engineering **728** (2020) 012002 doi:10.1088/1757-899X/728/1/012002

3.3 Results Comparison

In comparison, a table was created for individual values from the graphs, in which the average values of water absorption after 28 days were compared.

Table 1. Comparison of Volume Absorption of Polystyrene Samples with Different Surface Treatment

Wire cutting			Saw cutting			Original surface		
Average	32.320	%	Average	30.385	%	Average	27.154	%
Average deviation	0.702	%	Average deviation	0.412	%	Average deviation	0.58	%
Minimum	30.958	%	Minimum	29.598	%	Minimum	25.836	%
Max.	33.189	%	Max.	31.072	%	Max.	28.143	%

Table 2. Comparison of Weight Absorption of Polystyrene Samples with Different Surface Treatment

Wire cutting			Saw cutting			Original surface		
Average	94.196	%	Average	86.776	%	Average	83.153	%
Average deviation	0.898	%	Average deviation	0.905	%	Average deviation	0.429	%
Minimum	93.120	%	Minimum	85.818	%	Minimum	82.699	%
Max.	95.991	%	Max.	88.463	%	Max.	83.937	%

As we can see in the graphs, from the very beginning the individual samples differ in their course in terms of both volume and mass absorption. Over time, these differences increase, especially after 96 hours from the start of the measurement. However, due to the formation of inhomogeneous samples, it is not possible to interleave the measuring process with a function describing the absorption for expanded polystyrene only, due to the use of rubber-asphalt.

4. Conclusion

Surprisingly, a cutting wire appears to be the least suitable surface treatment of the expanded polystyrene with a closed outer structure. On the basis of comparison, it can be said that, especially in the weight capacity, this treatment becomes problematic. When compared to each other (see Table 2) the difference from the untreated original board surface after 28 days is up to 11%.

The surface treatment with a hand saw for cutting polystyrene is similar in its properties to the absorbent properties of the original untreated surface of the boards. Thus, the weight absorption, together with the volume absorption, is about 3.5% higher than the untreated original surface of the boards after 28 days. It can be noted, however, that the course of individual curves shows the greatest deviation compared to the other investigated adjustments with the highest average deviation from the tested variants.

This process is explained by the view of the microstructure of the individual surfaces after the different surface treatments.

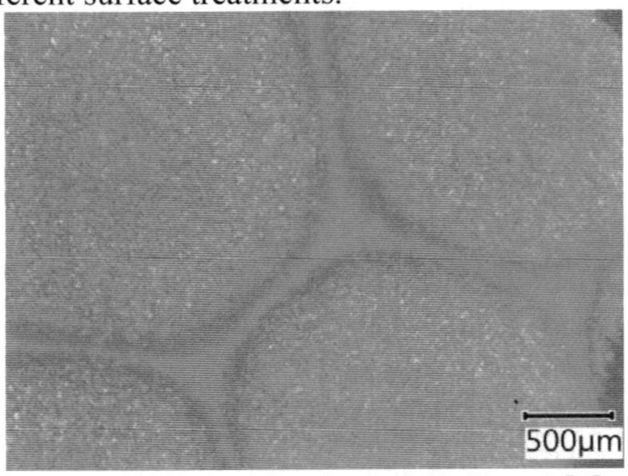

Figure 6. Original Surface of the Board [5].

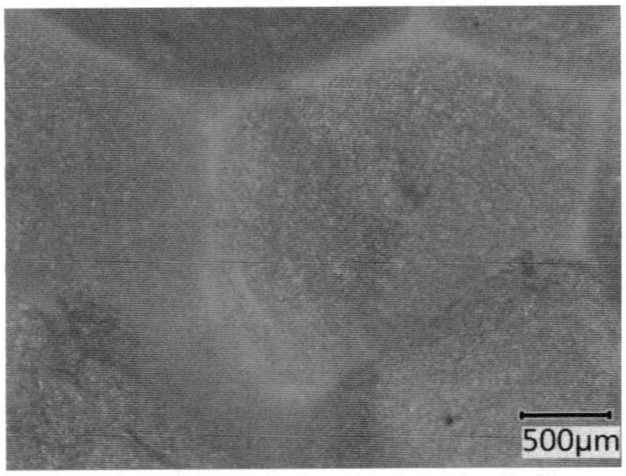

Figure 7. Saw Cutting of the Board [5].

Figure 8. Wire Cutting of the Board [5].

When the surfaces are approached 100 times, individual differences can be seen, which explain with great probability the completely different courses of absorption over time. As we can see in Figure 6 and Figure 7, the individual surfaces are very similar. While using the saw for cutting, minor mechanical damage to polystyrene beads can be observed. However, the damage does not occur on the microstructure and to the extent that would be decisive. Thus, the higher absorbency over the untreated original surface is very likely due to the larger area exposed to the adverse environment, ie moisture. It is also necessary to note the intact microstructures of the bead surface itself. The surfaces thus have a low porosity.

However, the surface in Figure 8, which has been treated with a cutting wire, shows great differences at first glance when compared to others. Individual beads of polystyrene were divided in different parts of their profile, whereby the microstructure was melted and thus significantly disturbed. When the surface is treated or cut with the cutting wire, the coating of the individual beads is melted and the remaining residue is then left to cool on the surface. This fact is best noticed on the capillary in the middle of Figure 8. In general, a surface with a much higher porosity and an unclosed structure is formed, which results in higher surface absorbency properties of the treated surface of the boards. Thus, it can be expected that this treatment will accelerate the deterioration of the heat-insulating properties, namely the coefficient of thermal conductivity, which is moisture-dependent [7].

It is therefore necessary in the further investigation to determine the long-term effect of repeated soaking of the surface, which is commonly found in the environment in which the material is placed. This investigation and research must be focused not only on the weight and volume absorption, but also on the speed of this absorption and, last but not least, on the thermo-technical and insulating properties.

Acknowledgement
This paper has been worked out under the project No. LO1408 "AdMaS UP - Advanced Materials, Structures and Technologies", supported by Ministry of Education, Youth and Sports under the „National Sustainability Programme I" and under the project No. FAST-S-19-6045 "Studium fyzikálních procesů v kritických detailech obalových konstrukcí budov s téměř nulovou potřebou energie".

IOP Conf. Series: Materials Science and Engineering **728** (2020) 012002 doi:10.1088/1757-899X/728/1/012002

References

[1] Directive 2010/31/EU of the European Parliament and of the Council of 19 May 2010 on the energy performance of buildings.
[2] TZB - Info – Tepelná izolace soklu a spodní stavby – chyby v praxi [online]. Accesible from: https://www.tzb-info.cz/4225-tepelna-izolace-soklu-a-spodni-stavby-chyby-v-praxi.
[3] ČSN EN 12087 Thermal insulating products for building application – Determination of long term water absorption by immersion, Praha ČNI, 2013.
[4] ČSN 72 2603 Testing of brickwork pruducts – Determination of mass, bulk density and water absorption, Praha ČNI, 1979.
[5] Author's Archive
[6] J. Plachy, J. Vysoka, R. Vejmelka, and Z. Caha, "Correlation of water absorption values of bitumen waterproofing sheets obtained according to CSN EN 14223 and CSN 503602," *Communications,* vol. 16, issue 4, pp. 118-122, 2014.
[7] A. Lakatos, "Influence of the moisture in the thermal conductivity of Expanded Polystyrene insulators,*" Contributions to Building Physics*, chapter 2.91, 2013.

IOP Conf. Series: Materials Science and Engineering **728** (2020) 012003 doi:10.1088/1757-899X/728/1/012003

Survey of the Designs of Wooden Architectural Monuments

E N Pokrovskaya [1]

[1]Moscow State University of Civil Engineering, 129337, 26, Yaroslavskoye Shosse, Moscow, Russia

feoktistovamgsu@mail.ru

Abstract. The article describes a method for examining the material of monument structures using physico-chemical studies. The surface of samples of wooden structures was investigated by electron microscopy. The study of the capillary structure of wood was determined by kinetic changes in water adsorption. The possibility of destruction of wood structures in time was determined by the mycological method. Wood samples were taken for examination from the following monuments of wooden architecture: The Kremlin of Rostov the Great; Kizhe Reserve Museum; St. Nicholas Church in the village of Llava; piles of the Assumption Cathedral STSL; Seraphim-Diveevo monastery.

1. Introduction

Monuments of wooden architecture are the cultural heritage of the world community [1-3]. Especially valuable with elegant unique architecture are the monuments of wooden architecture of Russia. To preserve the monuments of wooden architecture in time, it is necessary to periodically conduct a survey of structures. When restoring structures, it is necessary to determine the degree of their preservation and to predict their durability. To solve these problems, a physicochemical method was developed with the participation of mycological studies.

The main component of the polymer composite of wood is the ligno-carbohydrate complex (LUK). Cellulose and lignin are combined into a spatial polymer structure containing reactive hydroxyl groups. Preservation of wood strength is determined by the invariability of the chemical composition and structure of the onion. The presence of organic carbon-containing polymers in wood causes the exposure of wooden structures and wood products to biocorrosion. In the presence of biodegradable fungi, the destruction of wood occurs, which increases with wood moisture more than 18% [4]. For a reasonable determination of the durability of structures containing viable spores (FS) of biodegradable fungi, it is necessary to control the change by chemical analyses.

Cellulose and stucco (components of onion), as well as changing the structure of wood by electron microscopy. The strength of the samples depends on changes in the capillary structure of wood. Changing the capillary structure of wood affects strength and durability. This change was determined by the difference in water sorption.

2. Methodology

We had samples of designs of wooden architecture monuments from the Kremlin of Rostov the Great, the Kizhi Museum, a house on Moscow's Small Vasilyevsky Lane. The duration of operation of buildings and structures of these objects is 100, 150, 350, 400 and 1000 years. There were also wood samples from the pile foundations of the Assumption Cathedral of the Holy Trinity St. Sergius Lavra and piles of the Stables yard of the Rostov Kremlin; St. Nicholas Church from the village of Llava, Arkhangelsk Region, wooden beams on the floor of the Bishop's Hotel of the Seraphim-Diveevo Convent, etc.

Mycological studies were carried out to determine the type of biodegradors on wood samples and quantitatively determine viable spores by direct sowing of wood samples on solid agar nutrient media, followed by counting the number of viable spores of fungi per 1 sq.cm and determining the species composition of the community. Wood samples were placed on the surface of sterile wort agar and Saburo medium in Petri dishes and incubated at a temperature of 27-28°C and relative humidity of 90% for a week or more. The types of fungi were determined according to [5, 6], using the recommendations set forth in [4, 7].

As a physico-chemical studies, a chemical analysis of the wood composition was carried out for the Classon lignin content using the Komarov method, and cellulose was determined according to the Kürschner – Heffer method [8]. The method of IR spectroscopy was used to determine the chemical changes in the components of LAA [9]. The structure of wood was studied by electron microscopy [10].

3. Results and its discussion

The results of the mycological analysis of wood samples are shown in table 1. The kinetics of water sorption by the wood of the monuments made it possible to determine the changes associated with the specific surface of the sample material.

Table 1. The quantity of viable spores of fungi on wood samples.

No	Wood sampling location	Microorganisms (MO) detected in the sample		The number of FS per 1 sq. cm	The concentration of MO in 1 g of sample (total number)
		Wort Agar	Saburo Medium		
1	Seraphim-Diveevo Convent. Bishop's hotel. Wood floor beam	Ophistoma sp. Aspergillus niger; Aspergillus flavus; Penicillium cyaneo-fulvum; Penicillium biforme.	Ophistoma Aspergillus Penicillium Yeast Lypomyces	178	5·104
2	Kizhi. Seregin House	Aspergillus niger; Ophistoma; Aspergillus flavus; Fusarium; Serpula lacrimans	Serpula Actinomyces Lypomyces Yeast	126	4·103
3	Church of St. Nicholas, the village of Llava, from the eastern wall of the altar outside	Trichoderma viride; Trichoderma koningii; Cladosporium herbarum; Ophistoma	Penicillium Aspergillus Yeast Lypomyces Rodotoruba	214	9·105
4	STSL. Assumption Cathedral,	Aspergillus niger; Mucor hie ta Us Ophistoma, Fusarium	Yeast Lypomyces	193	2·107

	wooden foundation piles	Penicillium capculatum			
5	Maly Vlasyevsky Lane, Building 4. Part of the wall in the attic.	Penicillum biforme; Penicillum cyaneo-fulvum; Aspergillus glaucus; Aspergillus niger; Stemphylium sp;	Penicillum Yeast Lypomyces Candida Rhodotoruba	112	8·103
6	Piles of the Rostov Kremlin.	Penicillum chryso-genum; Penicillum capculatum; Aspergillus glaucus; Ophistoma; Fusarium	Penicillum Aspergillus Yeast	102	3·104
7	Rostov the Great. Stable yard. Depth 2.5 m. Pile.	Penicillum biforme; Cladosporium, Penicillum rugulosum, Fusarium	Penicillum Yeast Lypomyces	67	5·102

In the presented wood samples, mold fungi were revealed (5 species of Penicillum, 3 species of Aspergillus, Fusarium); biologically damaging - Trichoderma, Cladosporium, Mucor; wood-destroying Ophistoma serpula stemphylium; yeast and yeast-like fungi Candida, Lypomyces, Rodotoruba. The concentration of viable spores in the wood of damaged structures ranged from 67 to 214 sp./sq.cm. As is known from literary sources, the concentration of FS above 80 sp./sq.cm creates a dangerous situation [9]. The dependence of compressive strength along wood fibres on the number of viable spores is shown in Figure 1.

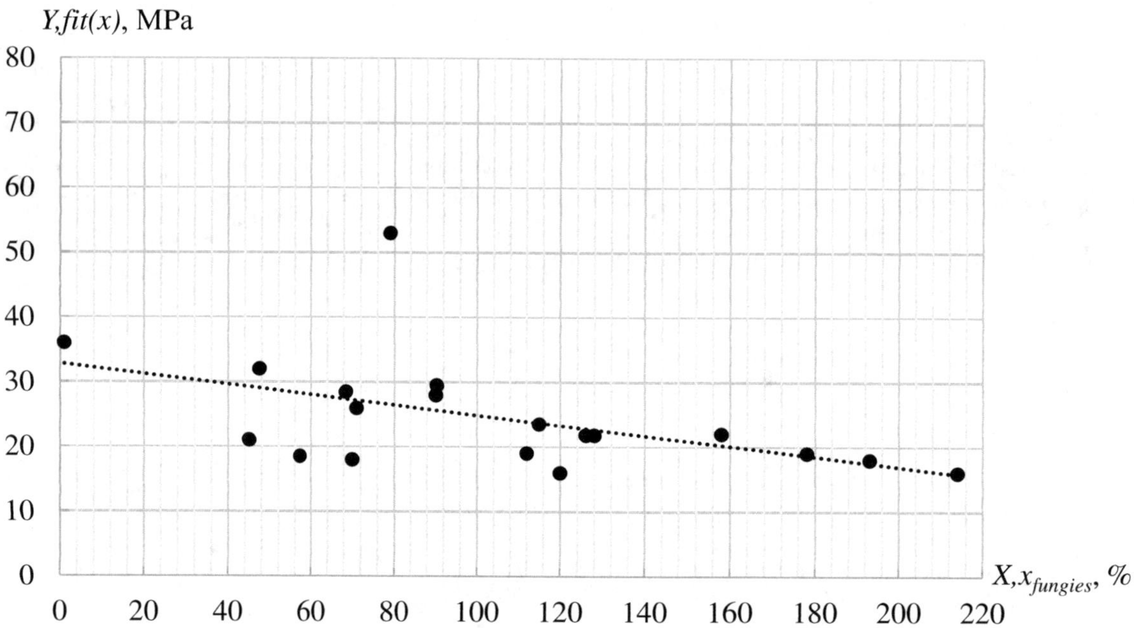

Figure 1. The dependence of the compressive strength along the wood fibres on the number of viable spores per sq.cm (the smallest strength of the samples is 1.3.4, table 1).

IOP Conf. Series: Materials Science and Engineering 728 (2020) 012003 doi:10.1088/1757-899X/728/1/012003

As shown by mycological studies, samples of the Bishop's hotel (floor beam) of the Seraphim-Diveevo monastery are more susceptible to biocorrosion. The outer wall (eastern) of the St. Nicholas Church, the village of Lavlya, wooden piles of the Assumption Cathedral STSL.

Wood adsorption of monuments of wooden architecture. Wood is a sorbent with a branched capillary-porous structure. Kinetic curves of sorption of woody water of various monuments are shown in Figure 2.

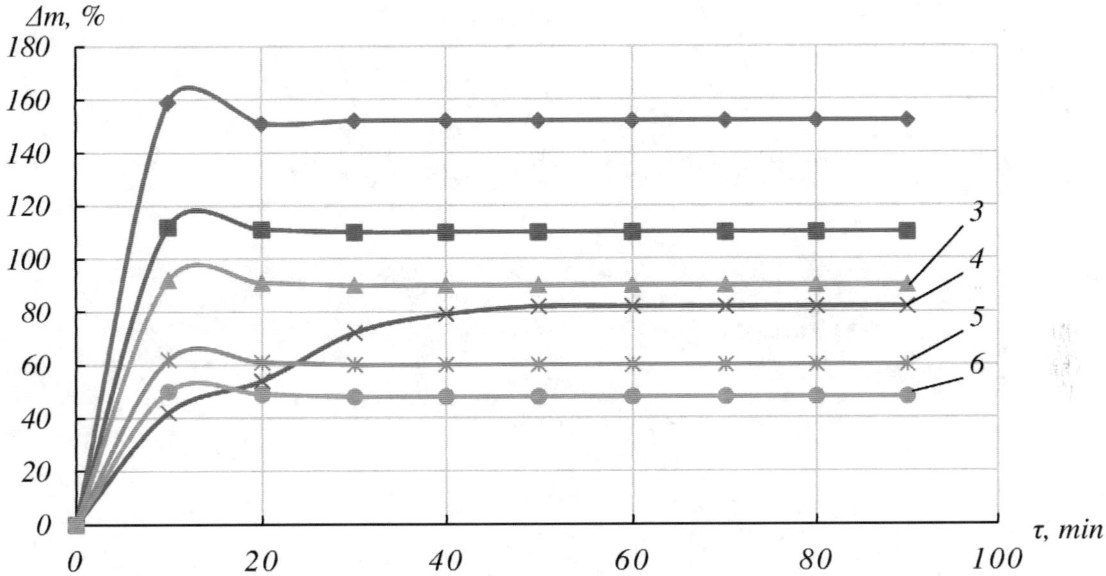

Figure 2. Kinetic water sorption curves of woody monuments of wooden architecture: 1 - source wood; 2 - eastern wall of the Nikolaev church; 3 - Bishop's hotel; 4 - Assumption Cathedral STSL; 5 - piles of the Rostov Kremlin; 6 - piles of the Rostov Kremlin (depth 2.5 m).

Structural failure (other than force impacts) begins from the surface. Destruction leads to "loosening" of the surface layer of wood [11]. Water absorption, water adsorption increases. One of the indicators of the destruction of the surface of a layer of a structure is the quantity A, the limiting value of adsorption. Of all the samples studied, the largest sorption of water is at the outer eastern wall of the St. Nicholas Church. (Figures 3-4).

Figure 3. The outer eastern wall of the St. Nicholas Church, Arkhangelsk region, Primorsky district, the village of Lavlya (1581-1584).

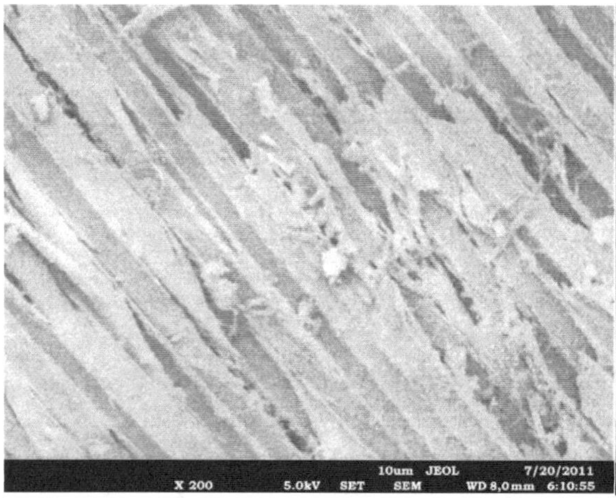

a)

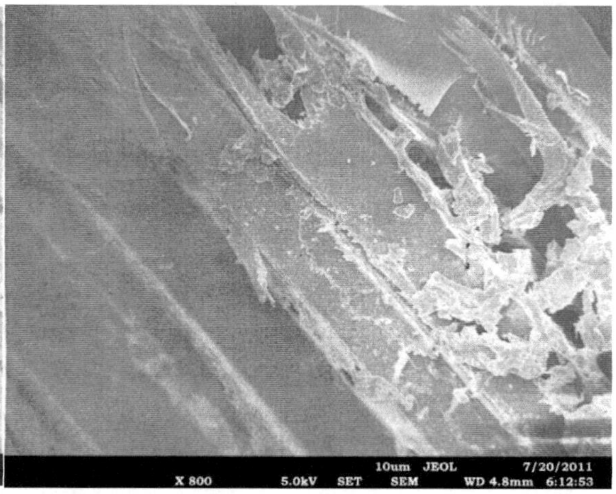

b)

c)

d

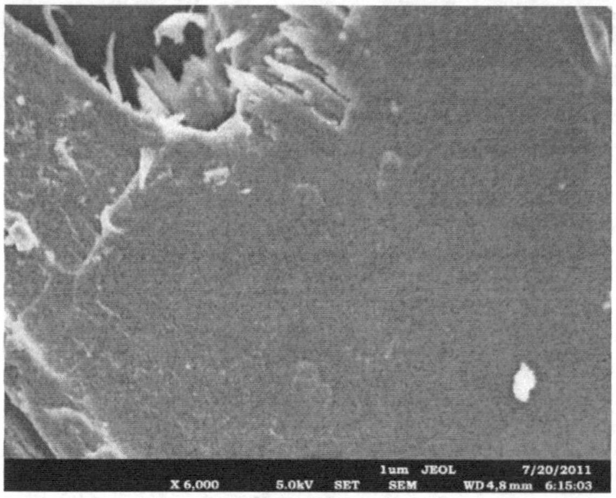

e)

Figure 4. The destruction of wood fibers on the eastern outer wall of the St. Nicholas Church (a - uv. 200x times; b - uv. 800x times); violation of the lignin matrix of the eastern outer wall of the St. Nicholas Church (c- in - at. 800x times; d - at. 2000x times; e - at. 6000x times).

4. Conclusions

The most objective method of examining wooden structures of monuments is physico-chemical studies and studies of mycological destruction of the material. Inspection of the designs of wooden architecture monuments showed once the largest internal change in the structure of wood samples: Nikolskaya Church, the village of Llava, from the eastern wall; wooden piles of the foundation of the Assumption Cathedral STSL; floor beams of the Seraphim-Diveevo nunnery, Bishop's hotel. During restoration, these objects should be treated with reinforcing antiseptic compounds, for example, the composition "Mipor".

References

[1] M. Matsuo, M. Yokoyama, K. Umemura, J. Sugiyama, S. Kawai, J. Gril., S. Kubodera, T. Mitsunani, H. Ozaki, M. Sakamoto, and M. Imamura, "Aging of wood: Analysis of color changes during natural aging and heat treatment," *Holzforschung*, vol. 65. Issue 3. pp. 361-368, 2011.

[2] H. Yorur, S. Kurt, and I. Ymrutas, "The effect of aging on various physical and mechanical properties of scotch pine wood used in construction of historical Safranbolu houses," *Drvna Industrija*, vol. 65, n. 3, pp. 191-196, 2014.

[3] M.L.-E. Florian, "Scope and history of archaeological wood," *Advances in Chemistry*, pp. 3-32, 1989.

[4] E. N. Pokrovskaya, *Physico-chemical basis for increasing the durability of wood*, Publishing house ASV, 2003.

[5] J. Perez, J. Munoz-Dorado, T. de la Rubia, and J. Martinez, "Biodegradation and biological treatments of cellulose, hemicellulose and lignin: an overview," *Int Microbiol.*, vol. 5, issue 2, pp. 53-63, 2002.

[6] P. Beguin, and JP. Aubert, "The biological degradation of cellulose," *FEMS Microbiol Rev.*, vol. 13, issue 1, pp. 25-28, 1994.

[7] I. G. Kanevskaya, *Biological damage to industrial materials*, 1984.

[8] *Biodeterioration and biocorrosion in construction: Materials of the International scientific and technical conference.* Saransk: Publishing House of Mordovia University, 2004.

[9] *A.* V. Obolenskaya, Z. P. Elnitskaya, and A. A. Leonovich, "Lab work on the chemistry of wood and cellulose," *Ecology*, 1991.

[10] *I.* N. Chistov, E. N. Pokrovskaya, "The study of wood historical monuments by infrared spectroscopy," *Vestnik MGSU,* Special Issue, no. 1. pp. 455-457, 2009.

[11] *E.* N. Pokrovskaya, Y. L. Kocalchuk, *Biocorrosion Preservation of historical and architectural monuments*, 2013.

IOP Conf. Series: Materials Science and Engineering **728** (2020) 012004 doi:10.1088/1757-899X/728/1/012004

Diagnostics of Flat Roofs with Flexible Sheets for Waterproofing

Stanislav Šutliak[1], Jan Plachý[2]

[1] Episs, s.r.o., Podjavorinskej 664, 027 43 Nižná, Slovakia
[2] Institute of technology and business in České Budějovice, Okružní 10, 37001 České Budějovice, Czech Republic

plachy@mail.vstecb.cz

Abstract. This paper deals with diagnostics of flat roofs with flexible sheets for waterproofing (membrane). The diagnostics is focused on defects and water tightness of membrane. The roof is one of the building's most important building envelope. In case of defect, the other layers of the roof, including the structure frame, are degraded. The introduction gives an overview of individual methods and their division. The following part describes advantages and drawbacks of the individual methods. The next part is focused on the following type of test - High Voltage Membrane Testing. The test is described in detail, including the Conductive detection membrane, which was developed specifically for this method. This test, in combination with other methods, seems optimal, but it is most reliable to prevent possible defects by following technological procedures.

1. Introduction

This paper deals with diagnostics of flat roofs with flexible sheets for waterproofing (membrane) with a focus on tightness of waterproofing. The issue is dealt with for example [1], electrical methods [2] and thermographic methods [3]. In the case of damage to the membrane, other layers of the roof could be severely degraded. This mainly concerns a possible degradation of the thermal-technical properties of the thermal insulation layer. The degradation is due to increased humidity.

Moisture on the inner surface may not always be a sign of damage of the waterproofing layer. Moisture on the inner surface may be caused by condensation. However, the amount of water (moisture) is higher in the case of leakage than in the case of condensation.

The diagnostics of flat roofs with layer waterproofing can be divided into a basic and a detailed research. The basic research uses visual checking and commonly available tools. The detailed research uses more sophisticate measuring tools and instruments. This paper focuses on these devices and methods.

As a part of the basic survey, the compliance with the application technology and the tightness of the membrane are checked. These are mainly laying membrane considering alternation of individual flexibile sheets, keeping the width of transverse and longitudinal overlaps, making joints (welding, gluing, roll-over) [4], mutual connection of layers for bitumen sheets and making details [5]. Checking the tightness of the membrane is aimed at checking the area and joints. In the case of an area inspection except of the joints, a visual control is applied to ensure that the membrane has not been damaged. In the case of polymer membrane with a signal layer, this control is easier. Polymer membranes which are composed of two layers have a bottom layer dark and a top light colour (yellow, orange). Joint testing needle is used to control the joints for the polymer membrane film. Pulling this Joint testing needle on the joint surface mechanically verifies the quality of the joint. The Joint testing needle should be slightly pushed on. As for bitumen sheets, specific instrument, so called an insulation spatula is used to check the joints. The procedure is the same as for polymer membrane films. Only for bitumen sheets there is the limitation for use at temperatures between 5 (10) - 20 (25) °C.

2. Material and methods of testing for detailed research

As mentioned above, more sophisticated instruments are used for detailed research. These devices are also used when the membrane is covered with another layer, for example, a roof with a vegetation layer.

Devices for diagnosis of flat roof defects can be divided according to destructive or non-destructive methods. Another division is based on whether the diagnosis directly identifies the defect or not - these are direct and indirect methods.

Direct method - the instruments which are used, can directly identify a mechanical defect in the membrane. The identification of this damage is either local (mostly in joints) or in the area except for joints. The manifestations of defect can then be recorded objectively (mostly air pressure measured) and the test is repeatable or indicative (usually we observe visually and, in this case, it is not possible to repeat the test). An overview of the methods, see the table No. 1., has been prepared on the basis of [6, 7].

Table 1. The quantity of non-combustible residues [6, 7].

Type of method	Testing method	Membrane damage	Tested area	Qualitative aspect
Direct	Press box	non-destructive	local	objective
	Press of double joints	destructive	local	objective
	Low voltage horizontal scanning platform	non-destructive	local	indicative
	Mapping vector of electric field with low voltage (vector mapping)	non-destructive	local	indicative
	Low voltage scanning of membrane surface	non-destructive	local	indicative
	Low voltage scanning of membrane surface	non-destructive	local	indicative
	Smoke method	destructive	whole area	objective
	Tracer-gas-box method	destructive	whole area	objective
	Overflow method	destructive	whole area	objective
Indirect	Impedant defectoscopy	non-destructive	local	objective
	Thermographic Defectoscopy	non-destructive	local	objective

2.1 Direct methods

2.1.1 *Pressure box*

The test is based on the application of a transparent bell on the membrane surface, which is covered with a solution of water and soap or detergent. Vacuum of about 20 kPa is created under the bell. In case of a damage of insulation there is leak and bubbles appear on the surface. The disadvantage of the method is time and work consuming procedure. Also, there is the temperature limitation of use od this method. Due to the solution, the method cannot be carried out at temperatures around 0°C and bellow. The method is mainly used for polymer membranes.

2.1.2 *Double joints press*
This test can be performed for double and folded joints. These joints are mainly made with polymeric films. The test is based on creating overpressure inside the joint. The pressure must not drop by 10% during the first 10 minutes. See No. [1].

2.1.3 Low Voltage Horizontal Membrane Scanning Platform.
When using low-voltage electrical potential scanning, the membrane must be moistened (water-covered). Moisture gets under the membrane at the point of breach. For methods using low voltage electrical potential, a controlled covering of water on the surface forms the conductive path horizontally across the membrane to any membrane breach. At a breach location, an electrical path to the deck is formed through the water leaking to the deck below. A sensitive receiver detects the leakage current and alerts the operator. For further details, see No. [2]. Membrane Scanning Platform is based on creating voltage potential between platform (membrane) and roof deck and detecting any leakage current passing through the membrane. After the tested membrane is wetted, a voltage with respect to area is generated, followed by locating the areas where electrical current flows from the platform through membrane breaches to the deck. The platform has two sets of metal sweeps making continuous electrical contact with the membrane surface. The outer sweep forms a continuous perimeter around the platform with the inner sweep contained within the perimeter of the outer sweep.

2.1.4 Low Voltage Membrane Electric Field Vector Mapping
The principle is as described in [2]: "The electric field vector mapping technique uses an electric potential gradient across the membrane surface, along with a sensitive voltmeter and probes to locate membrane leaks. Conductor cable loop is installed around the perimeter of the tested area. A signal generator is connected to the loop cable and the building ground. The area within the loop is sprayed with water to make a continuous conductive surface in the tested area. Most roofing/waterproofing membranes are nonconductive. Thus, the electrical signal from the perimeter cable loop, searching for an electrical path over the wet area of the roof to breach area, completes the circuit to the substrate. The resulting current from the point of breach to the perimeter cable sets up a voltage gradient in the water within the perimeter."

2.1.5 Low Voltage Vertical Membrane Surface Scanning
The principle is as described in [2]: "The moistened sensor, which is connected with a cable to the voltage source in the receiver, is pressed onto the tested surface. This pushes the water onto the membrane surface. A leakage current then flows over the moistened sensor. Consequently, the receiver registers a deflection on the signal level meter accompanied by an audible alert. The vertical surface leak locator is used for testing the membrane integrity of corners, parapet walls, and seams."

2.1.6 High Voltage Membrane Testing (spark test)
The principle is as described in [2]: "High voltage testing is performed on a dry horizontal or vertical surface using a limited current at voltage within 10 - 40 kV, depending on the membrane thickness. One conductor from the portable current generator is grounded to the roof deck with a sufficient conductive substrate. Another conductor is attached to a special electrode brush made of conductive metal bristles. The brush electrode is then swept over the surface of the roof membrane. An electric arc jumps from the electrode when any damage in the membrane, completing a circuit between the brush and the roof deck. Where there are no damages, the membrane acts as an insulator and prevents the flow of current to the deck. This method may be used to test waterproofing up to the thickness 26 mm. For the effective test, there must be conductive layer under membrane."

2.1.7 Smoke method
Leakiness of the membrane is tested by driving of thick or coloured smoke under the membrane. This test is suitable for one - layer systems. Construction frame must be airtight, or must consist of vapour-

barrier layer. Observed smoke reveals the point of breach. The pipes are part of the roof system; thus, it is necessary to damage membrane and consequently to locally repair it. The method is quick, however, there is the limitation of use as for the size of defect. Based on our experience, the revealed area must exceed the size of 10 mm. For more details, see [1].

2.1.8 Tracer-gas-box method
The test is similar to Smoke method. Leak tightness of the membrane is proved by detecting gas on the nitrogen and hydrogen basis. Hydrogen is recommended to check the leakiness. It has the highest molecular speed and its viscosity is lower than of other gases. However, it is necessary to use its nitrogen form.

2.1.9 Overflow method
In this test the membrane surface is covered by continuous water layer and drop of water level is observed. The test is very risky. It is necessary to consider using the test while designing new construction. It is not suitable for the finished roofs, as there is the risk of damaging the frame construction in case of leaking under the membrane and damaging the waste water pipes. The test is time - consuming. Minimum length of the test is 48 h. For more details, see No. [1].

2.2 Indirect methods - Humidity-detection method

2.2.2 Impedant defectoscopy
For this method special electronic resistance sensor is used. The principle of electric impedance enables the signal to penetrate the depth up to 100 mm. The signal detects the moisture even in the base layers under the membrane. There are some weather conditions limitations for this method. The temperature must not decrease bellow +4°C, and there must be no rain or increased air humidity approx. 12hours before starting the test. It is not possible to use this method for testing hydro insulation that include conductive element. (e.g. bitumen sheet with Aluminium foil), and covered with another layer (a concrete paver, gravel, substrate for green roofs).

2.2.3 Thermografical defectoscopy
The method is based on temperature gradient of dry and wet area. The dry area is cooled down and heated slower than wet area. At the sunset or shortly after that the dry areas are cooled faster. The wet area, thus have higher temperature. For the determined experiment conditions see No. [3]. They are the minimum inside–outside temperature difference, minimum daytime–night-time temperature swings, maximum wind speed, roof conditions, and precipitation limitations within 24 hours leading up to the scan. Evaluation of the results of infra-red data requires a large experience.

3. Results
The analysis of the methods results in the conclusion that there is no universal test for different roof constructions.

If checking the roof already leaking, it is possible to assume moisture in the roof construction. In this case, the combination of impedant defectoscopy along with some of the electric methods seems optimal.

The method is necessary to be considered already when designing the roof construction. There are two solutions: non-conductive construction, b - conductive construction.

The new non-conductive construction is possible to check only after exposure to water stress. Consequently, the roof acts like the leaked roof. As a result, it is possible to use the two methods

IOP Conf. Series: Materials Science and Engineering **728** (2020) 012004 doi:10.1088/1757-899X/728/1/012004

mentioned above (the combination of impedant defectoscopy along with some of the electric methods). In short, only the wet breach is found.

The new conductive construction is tested without exposing to water stress, as the water may cause large damage when leaking the construction. Conductive construction may be tested by some of the electric methods. There is the advantage of higher accuracy and possibility of detecting a potential leaking point already before leaking. In short, the dry hole is detected, which may have appeared just before testing. Conductive constructions, however, cannot be tested by impedant defectoscopy.

Another problem may have reconstructed roofs with polymer membrane without thermal insulation. New polymer membrane is necessary to separate from the current membranes. The occurred condensation between the new and old membrane during the year influences the results of impedance defectoscopy in negative way. The conductive layer under membrane thus must perform separating function. As a result, conductive layer has been developed performing the terms mentioned above. Se the picture No. 1.

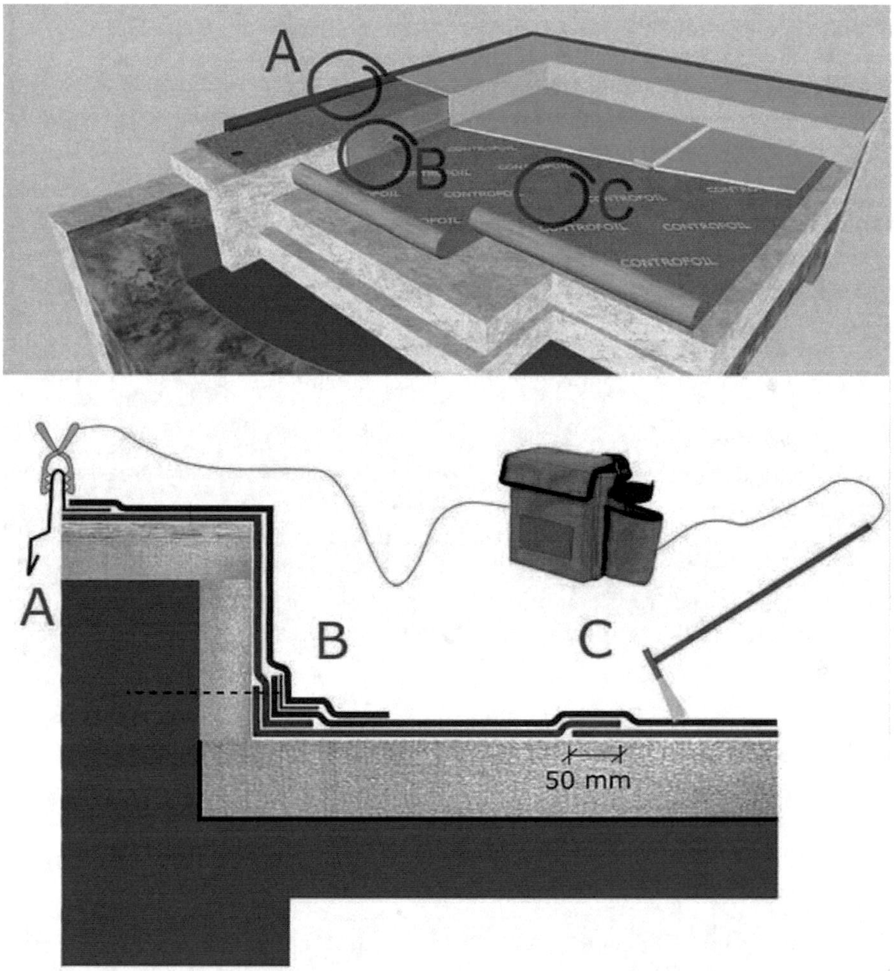

Figure 1. Position of conductive membrane under waterproofing [8].

IOP Conf. Series: Materials Science and Engineering **728** (2020) 012004 doi:10.1088/1757-899X/728/1/012004

4. Discussion

On the basis of the research the conductive detection membrane with separating function has been developed. Conductive layer is multilayer reinforced with perforated aluminium foil.

The upper layer consists of the blue coating and perforated aluminium foil. Due to perforation, S_d = 0,65 m. The temperature conditions for the use of conductive foil are determined within -30/+ 80°C.

The application of the conductive detection membrane is as follows. Conductive detection membrane is placed on the thermal insulation material directly under waterproof membrane (SBS bitumen, PVC-P membrane). By overlapping at least 50 mm creates a full-surface conductive layer.

- In case of roofs with the overburden it is not necessary to use geotextile between materials such as: EPS polystyrene and PVC-P membrane, because of incompatibility, as the conductive and separating layer has suitable properties,
- the conductive layer - conductive detection membrane is finished on a horizontal surface of the parapet wall under a PVC coated metal sheet (drip mould) and creates a circumferential anchor point for performing a high voltage testing.

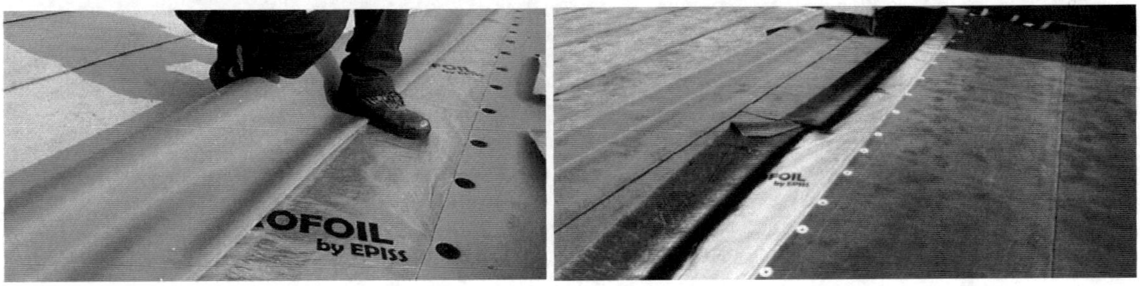

A B

Figure 2. Application of conductive detection foil under waterproofing. A - waterproofing from PVC-P, B - waterproofing from bitumen sheets [9].

5. Conclusion

Electric methods are commonly used for Diagnostic of the waterproofing membranes. Particular attention should be paid to testing single-layer membrane systems [10]. High voltage membrane system testing and other electric methods require the completed electrical circuit before detecting the membrane breach, both above and under the membrane. To accomplish the closed electrical circuit for the roofs with classic layer order, there must be placed the conductive layer or conductive material directly under the tested membrane.

References

[1] J. Pešta, D. Tesař, and V. Zweiner. *Diagnostika staveb. hydroizolace, termografie, blower door test, akustika,* 2014 (In Czech).

[2] ASTM D7877-14, Standard Guide for Electronic Methods for Detecting and Locating Leaks in Waterproof Membranes, ASTM International, West Conshohocken, PA, 2014.

[3] ASTM C1153-10(2015), Standard Practice for Location of Wet Insulation in Roofing Systems Using Infrared Imaging, ASTM International, West Conshohocken, PA, 2015.

[4] T. Petricek, P. Kacalek, "Dependence between the width of strip overlap and the strength of joints of one-layer bitumen sheets," *MATEC Web of Conferences*, vol. 146, 02007, 2018.

[5] R. Smolka, R. Sobotka, and J. Šulistova, "Design proposal for door thresholds," *MATEC Web of Conferences*, vol. 93, 01009, 2016.

[6] Company archive Episs, s.r.o.
[7] Personal archive of author Jan Plachý (Unpublished).
[8] Episs, s.r.o. 2019. *Iskrova skuska.* [cit. 2019-09-27]. Accesible from: https://www.episs.sk/iskrova-skuska (In Slovak).
[9] Episs, s.r.o. 2019. *Controfoil.* [cit. 2019-09-27]. Accesible from: https://www.episs.sk/controfoil (In Slovak).
[10] T. Petricek, P. Kacalek, and R. Smolka, "Single-layer mechanical fastening waterproofing systems," *Advance Materials Research*, vol. 1041, pp.71 -74, 2014.

IOP Conf. Series: Materials Science and Engineering **728** (2020) 012005 doi:10.1088/1757-899X/728/1/012005

Secondary Raw Materials in Construction

Radim Smolka [1], Jindřich Sobotka [1]

[1] Brno University of Technology, Faculty of Civil Engineering, Institute of Building Structures, Veveří 95, 602 00 Brno, Czech Republic

smolka.r@fce.vutbr.cz

Abstract. The paper describes and evaluates the use of secondary materials in construction details of buildings, primarily in the area of flat roofs. The roofs have a large number of technologies that are necessary for the smooth and healthy use of buildings according to current standards. We are currently discussing the take-back of PET bottles in the Czech Republic. New materials must meet specified parameters. Not only thermal, but also hygienic and fire-fighting. Primarily failures of roof structures are problematic parts during repairs.

1. Introduction

The Czech Republic sometimes takes over the standards of neigh boring countries. However, differences in climatic conditions can be quite crucial. The need for own standards is therefore evident. In this respect, it can also be implemented in the concept of recycling and in civil engineering as a whole. Very frequent wastes are from demolition of unsatisfactory buildings, asphalt wastes and plastics in large quantities. Generated municipal waste in large cities is sorted and unsorted, primarily as glass, plastic, paper and electrical components. From a logical point of view, the polymeric secondary material is unsuitable for primary use in the food industry. It is already hygienically defective and cannot be fully cleaned. However, it is suitable to use the polymer in the construction industry, for example as substructure for flat roof technology. It creates flat surfaces that do not prevent water runoff and do not obstruct roof maintenance.

2. Potential use of recycled materials in construction details

On flat roofs it becomes necessary to apply technological equipment to a large extent. The location in the upper part of the building is offered from several points of view - spatial, financial and operational. The secondary requirements for roofing layers, such as the maintenance of technological equipment and the safety of movement on the roof, need to be evaluated.

Primary requirements on elements:

- long life span,
- frost resistance,
- constant volume,
- tolerance of other materials (material corrosion),
- strength – resistance to impacts and constant load,
- fast and all-year-round construction without technological changes,
- meeting requirements for fire resistance.

The above parameters are necessary for good function and durability of the inserted element.

2.1 Production and testing of testing samples
2.1.1 Production of testing samples
Test flat plate elements of 150 x 150 x 15 mm were produced from waste HDPE [1]. Other elements such as sawdust and blowing agents have been produced, tested and measured to primarily improve thermal, technical and mechanical properties. After measuring the coefficient of heat penetration, the boards were tested for peeling at connections [2, 3].

Figure 1. Testing board – HDPE with added sawdust.

Four elements of bitumen sheets were applied on the testing board by fusing. Two on each side.

2.1.2 Resistance from peeling at connections
The test was performed in a laboratory of company Dehtochema bitumat-tn a.s. The testing method is designed for testing connections in mechanically fixed one-layer bitumen roofing. Technical Definition - resistance to peeling is a tensile force necessary to complete separation of prepared connections. During testing, the testing body is pulled by a constant speed up to its complete separation.

The test was performed with the use of a testing tearing device LABORTECH 2.050, which is shown in Figure 2. The device is equipped with continuous recording of force which corresponds with the distance of grips and it is able to keep balanced speed of separating the grips. The grip width may not be lower than 50 mm. The speed of separating is defined as 100 mm ± 10 mm per minute. The testing body is secured against displacement by an inserted stopper.

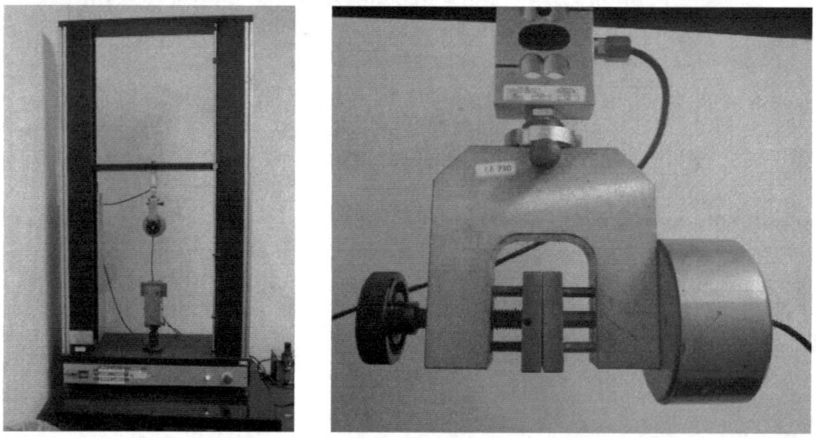

Figure 2. Testing tearing device Labortech 2.050 and a detail of its upper grips.

The testing samples for connections were tempered (left) for 48 hours at the temperature of 22°C – 24°C under relative humidity of 60 %. The tempering conditions met the standard conditions as well as the necessary standard conditions [4,5], i.e. 20 hours at the temperature (23 ± 2) °C and relative humidity of (50 ± 20) %.

The measured values showed the possible use of secondary plastic raw materials in the construction details of buildings, in this case flat roofs. The tests were also positive for use with a common type of waterproofing. Physical, mechanical, and chemical properties puts them in polymer-compositions where traditional heat insulation materials are used. Using a product from secondary materials results in a lower thickness of a structure, since no other waterproofing or spreading layers are necessary, while the requirements of technological standards, regulations, and legislation acts are met. The big advantage is theoretically lower financial burden.

3. Classification of current critical details

At present, the base layer is laid directly on the waterproofing layer. (Figure 3a, 3b). The concrete distribution block is used as a suitable base distribution element, but it constitutes an obstacle to the natural runoff. It is therefore very difficult to create scattering wedges and to drain water from such inaccessible areas of the section blocks and technological systems.

Figure 3a, 3b. Pushing of distributing concrete structure to roof cladding.

In order to avoid the above-mentioned errors and imperfections, it is possible to use a thermal insulator which has higher thermal performance than the polymer but has a lower compressive strength. In this case, however, there is a possible additional failure of the waterproofing layer, which cannot withstand the pressure increase and a depression is formed.

The figures (Figures 4a, 4b) above illustrate a relatively common phenomenon of an improperly selected distribution structure below the process equipment. Rainwater settles at the point of extrusion and does not flow smoothly from the roof structure, which may result in gradual degradation of the waterproofing layer. A similar phenomenon also occurs at a point load from the technology support structure.

Figure 4a, 4b. Technological device on band supports and detail support.

Previously, it was not considered in the design with the location of technology, such as the installation of air conditioning and air conditioning units. The roofs were therefore not primarily intended for the installation of fixtures.

4. Application of plastic in flat roofs

This chapter discusses the possible use of secondary plastics in the construction details of buildings. The test sample is designed and produced is considered to be placed on the surface as well as at the end detail, e.g. at the attic. The connection of the element with the waterproofing layer is necessary for its proper functioning.

On the basis of measurements and tested, primary points have been defined that must be met within the framework of the introduction of the element into the field of building construction. The product should be able to withstand heavy loads, be heat insulating, be waterproof (secondary), allow easy and smooth installation and should not suffer from material corrosion. All newly designed buildings can therefore be equipped with the necessary technology either in the design or in the future.

The following basic properties the element should have are considered:

- long life span,
- constant volume,
- tolerance of other materials (material corrosion),
- strength – resistance to impacts and constant load,
- fast and all-year-round construction without technological changes,
- meeting requirements for fire resistance,
- harmless to health,
- sound insulation.

The above-mentioned characteristics are necessary for correct and long-term function and long life of the designed product [6-9].

5. Conclusion

Designed base plate that summarizes these important parameters
was subjected to measurements peeling in joints confirmed placement of plastic boards in construction details of roof structures. Checking:

a) potential local joining,
b) potential all-area setting.

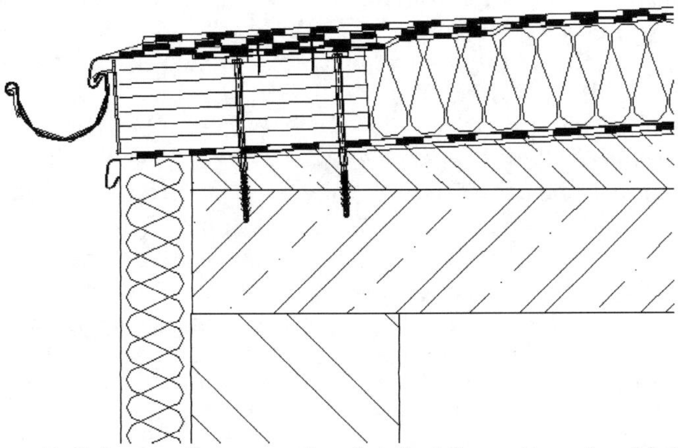

Figure 5. Scheme of construction detail at the gutter – local joining.

Horizontal joining of boards (Figure 5) is suitable for roof construction details, their locks are multipurpose and can be used as corners at the attic, at the end of the structure at the light well, at roof hatches, etc. Thus, the proposed board can very well replace standard thermal insulators such as XPS.

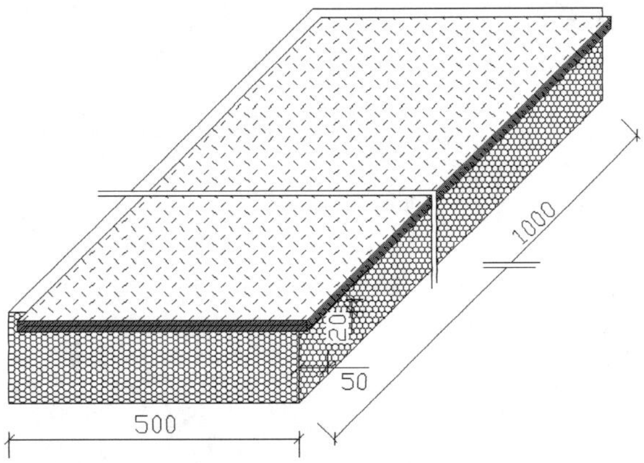

Figure 6. Designed board for full-area setting.

A polymeric element which may be provided with an integrated thermal insulator layer at the factory (Figure 6) is designed for larger areas. The product already has overlaps of 50 mm. These overhangs serve as locks and thus continuously transmit the load on the surface to adjacent plates. The products are therefore suitable for roofing as a distribution layer.

Acknowledgment(s)
This paper has been produced within the project FAST-J-18-5506 - Study of drainage factors of retention roof compositions using recycled polyester boards

References

[1] L. Matejka, and R. Smolka, *Support structures for a doors threshold. EP 2159364 B1, patent. Munchen, 2012.*
[2] V. Duchacek, *Polymers – production, properties, processing, applications*, 2006.
[3] ČSN EN 12316-1 (727638). *Flexible sheets for waterproofing - Part 1: Bitumen sheets for roof waterproofing - Determination of peel resistance of joints*. Praha: Úřad pro technickou normalizaci, metrologii, a státní zkušebnictví, 11/2000.

[4] R. Smolka., and J. Sobotka, "Application of Recycled Plastic in Flat Roofs," *MATEC Web of Conferences,* vol. 146, 02009, 2018.

[5] J. Plachý, "The problem of the compatibility of bitumen sheets for the reconstruction and rehabilitation of roofs," *MATEC Web of Conferences*, vol. 93, 02004, 2017.

[6] P. Kacálek, R. Smolka, and T. Petříček, "Design Properties of Log Walls with Respect to Building Physics," *Advanced Materials Research*, vol. 1041, pp. 288-292, 2014.

[7] T. Petříček, P. Kacálek, and R. Smolka, "Single-Layer Mechanical Fastening Waterproofing Systems," *Advanced Materials Research*, vol. 1041, pp. 71-74, 2014.

[8] T. Petříček, and P. Kacálek, "Dependence between the width of strip overlap and the strength of joints of one-layer bitumen sheets," *MATEC Web of Conferences,* vol. 146, 02007, 2018.

[9] J. Sobotka, and R. Smolka, "The temperature distribution during the microwave drying of wooden elements for structural defects in buildings," *MATEC Web of Conferences, vol. 146,* 03008, 2018.

Surface Temperature of Flat Roofs with Waterproofing Polymer Membranes

Jan Plachý[1], Jana Vysoká[1]

[1] Institute of technology and business in České Budějovice, Okružní 10, 37001 České Budějovice, the Czech Republic

plachy@mail.vstecb.cz

Abstract. This paper deals with the problematics of the surface temperature of flat roofs, where the waterproofing layer is used as polymeric waterproofing membrane of softened PVC. Currently, it is possible to encounter defects on flat roofs due to the high surface temperature of the waterproofing layer. The consequence of such a high surface temperature is, for example, the sublimation of the thermal insulation of the foam polystyrene. The paper focuses on the influence of the color of the plasticized PVC membrane on the surface temperature. The aim of this paper is to determine the maximum surface temperature of a polymeric PVC membranes in the dependence on color on the basis of experimental measurements and furthermore, to determine the maximum temperature on the surface of the thermal insulation under PVC polymeric membrane. Measurements were taken during a warm summer day. Used samples were represented by polymeric membranes in green, gray, red and brown colors. Mineral wool was used as a substrate under the polymeric membrane.

1. Introduction

This paper is focused on the analysis of the surface temperature of waterproofing polymeric membranes made of soft PVC in the compositions of flat roofs with the classical order of layers. In compositions with the classical order of layers, where the waterproofing is above the thermal insulation layer, the polymeric membranes are exposed directly to the sunlight. One waterproofing layer systems, either bitumen sheets or polymer membrane, can be damaged easily [1-2].

The high surface temperature is one of the causes of failure on flat roofs with coated waterproofing. These failures are manifested, for example, by sublimation of the thermal insulation of the foam polystyrene. Previously it has been dealt with the problem of surface temperatures on polymeric membranes in [3]. As a result, is the following finding: the temperature of the thermal insulation surface in the combination with the reflection of solar radiation exceeds 80°C for a long time.

The surface temperature of polymeric membranes is mainly influenced by [4]:

1. Surface properties of bitumen membranes (emissivity, reflection),
2. Substrate (thermal capacity),
3. Ambient (air) temperature;
4. Radiation in surroundings

5. Air flow and humidity,
6. Position in relation to cardinal points;
7. Waterproofing position (a slope in relation to the horizontal position).

In view of the surface properties of the polymeric membranes, the choice of the surface color is important for the solar reflectance and emissivity of the surface.

The aim of this paper is to find out what maximum temperatures are achieved on the surface of polymeric membranes and on the surface of thermal insulation in terms of color of the membrane without the influence of the reflection and whether this temperature approaches the limit of 80°C. This value is the limit guaranteed by manufacturers of foam polystyrene for its stability [5]. Thus, it is possible to establish the hypothesis that the surface temperature of the polymeric membranes and the thermal insulation surface does not exceed the value of 80 °C. Another hypothesis that can be put forward is as follows: Does the temperature range between the maximum temperatures of different colors reach values of more than 10°C? The last hypothesis would be formulated as follows: will be the lowest temperature achieved in light colors and the highest temperature will be achieved in dark colors.

Nowadays, when temperatures rise every year in the summer months, this issue becomes very actual. The measurements were carried out during the summer of the year 2018 in the range from 12.8.2018 to 14.8.2018. In the long term, the summer 2018 can be considered the second warmest summer in the 245-year history of temperature monitoring in the Czech Republic [6].

2. Material and methods of testing

2.1 Material

2.1.1 Samples of polymeric membranes

For the experimental part, polymeric membranes of softened PVC of the thickness 1.6 mm with a surface texture were chosen. As to the the color selection, light gray, dark gray, red, brown, light green, dark green and gray samples were used, see Figure 1. The 100 x 50 mm polymeric membrane samples were laid loosely on a mineral wool substrate (λ_D = 0.037 W/mK) of thickness 100 mm in special wooden frame. The sample frame lay on a flat roof. The samples were directly exposed to the sunlight without further reflection of the sun rays.

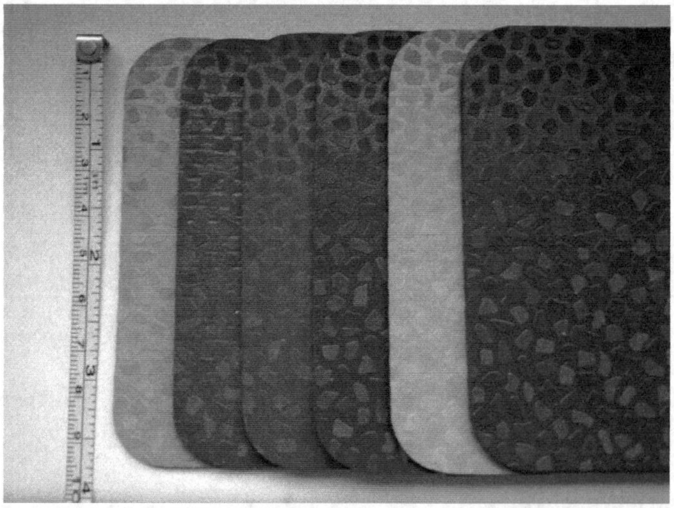

Figure 1. Polymeric PVC membrane. A – light grey, B – dark grey, C – red, D – brown, E – light green, F – dark green.

2.1.2 Apparatus
Two methods were used for the temperature measurement – a contactless and a contact method. The contactless method measures the temperature on the surface of the polymeric membrane and also contains the reflected component of the solar radiation. For measurements by the contactless method was used IR thermometer Voltcraft IR-650-12D with the temperature resolution of the value 0.1°C and with the basic accuracy of ± 3°C. The set emissivity of the value 0.92 corresponded to the surface behaviour of the polymeric membranes in the IR spectrum. The emissivity value of common building materials ranges from 0.85 to 0.95.

The contact method measures the temperature on the lower surface of the polymeric membrane, i.e. the upper surface of the thermal insulation, by means of a thermocouple. The universal measuring device ALMEMO 2590-4 with a thermocouple TYPE K - T190-1 with a temperature resolution of 0.1°C and a measurement accuracy of 0.03% of the unmeasured value was used for the contact method.

2.2 Methods of testing
Samples of polymeric membranes were placed on a flat roof with a 5° inclination oriented to the southeast. In terms of air flow, the roof is located in a normal landscape, on a protected, detached building at a height of about 2.0 m above the terrain at an altitude of about 425 m above sea level in the village called Planá nad Lužnicí (GPS: 49.3630811N, 14.7196306E). The air temperature was measured about 15 mm above the mineral wool substrate using a TyP K thermocouple. A low aperture was created around the samples to eliminate the effects of wind [4].

3. Results
Results of the measurements using both methods are given in Table 1. The maximum surface temperatures were recorded at the air temperature in 13.8.2018 at 42.6°C. The maximum temperature at the measuring station of the Czech Hydrometerological Institute (CHMU) on this day in the Czech Republic was measured in Radovesnice at 35.7°C [7]. It was the 7th - 9th hottest day of summer 2018 [7].

Table 1. Overview of measured maximum temperature values of polymeric membranes in dependence on surface color.

Color	Contactless method (13.8.2018) [°C]	Contact method (13.8.2018) [°C]
light grey	70.2	68.8
dark grey	80.3	71.0
red	77.0	71.9
brown	79.7	70.6
light green	70.4	69.1
dark green	77.3	69.6

4. Discussion
The maximum temperature of 80°C was reached on a single surface during a non-contact measurement. The hypothesis was therefore only partially confirmed. In the case of the contactless method, temperatures from 68.8°C to 71.9°C were reached. Thus, the temperature range was 3.1°C. Temperatures of 70.2°C to 80.3°C were achieved by using the contact method. Thus, the temperature range was 10.0°C. Thus, a hypothesis on a temperature range of more than 10°C was confirmed.

The temperature difference in the group of samples of various colors for the non-contact and contact method is significantly lower than in the case of bitumen membranes [4]. This small difference, in the case of non-contact method (only 3.1°C), is due to the low specific capacity of the thin polymeric membrane. Large differences when using the contactless method can be attributed to the reflected component of the solar radiation. In all cases, the temperature on the surface of the thermal insulation was lower than on the surface of the polymeric membrane. Contrary to the assumption, for all materials, there was a little difference between the non-contact and contact measurement method for the same sample. It was confirmed for both methods that the highest temperatures were reached for samples in dark colors. The hypothesis was therefore confirmed.

5. Conclusion

The obtained results show that the color of the polymeric membrane has a significant effect on the surface temperature and thus the acceleration of the aging process. Although the assumption of reaching the temperature of 80°C on the surface of the thermal insulation has not been confirmed, light colors should be preferred over dark colors.

In the next phase of the research we would like focus on the measurement of surface temperatures on thermal insulation of foam polystyrene (EPS). In the case of this type of thermal insulation, a separation layer is sandwiched between the soften PVC membrane and the thermal insulation. The separation layer is a high basis weight geotextile and may affect measurement results.

References

[1] T. Petricek, P. Kacalek, and R. Smolka, "Single-layer mechanical fastening waterproofing systems," *Advance Materials Research*, vol. 1041, pp.71 -74, 2014.

[2] T. Petricek, and P. Kacalek, "Dependence between the width of strip overlap and the strength of joints of one-layer bitumen sheets," *MATEC Web of Conferences*, vol. 146, 02007, 2018.

[3] O. Zidek, and A. Zak, "Měření teplot ovlivněných odrazem slunečního záření od prosklených ploch v polystyrenu na střeše s povlakovou hydroizolací". (DEKTRADE, Praha, 2017). [cit. 2019 -09-20]. Accesible from: https://www.dekpartner.cz/aktuality/detail/224 (In Czech).

[4] J. Plachy, J. Vysoka and R.Vejmelka, "Surface temperature of bitumen sheets in the flat roof structure, " *MATEC Web of Conferences*, vol. 279, 02017, 2019.

[5] K. Chaloupka, and P. Rydlo. *Izolační praxe 3. Ploché střechy a pěnový polystyren*. Sdružení EPS ČR, (2016). [cit. 2019 -09-20]. Accesible from: http://www.epscr.cz/obj/1145/IP_3_Ploche_strechy_a_penovy_polystyren.pdf (In Czech).

[6] CHMU, 2019 [cit. 2019 -09-20]. Accesible from: http://www.infomet.cz/index.php?id=read&idd= 1567755441 (In Czech).

[7] CHMU, 2019 [cit. 2019 -09-20]. Accesible from: http://portal.chmi.cz/files/portal/docs/poboc/OS/OMK/mapy/prohlizec.html?map=TMA (In Czech).

Heating and Drying of Autoclaved Aerated Concrete Blocks for Defects of Buildings by Application of Microwave Radiation

Jindřich Sobotka [1], Radim Smolka [1]

[1] Brno University of Technology, Faculty of Civil Engineering,
Institute of Building Structures, Veveří 95, 602 00 Brno, Czech Republic

Sobotka.j@fce.vutbr.cz

Abstract. Experiments of heating and drying of autoclaved aerated concrete blocks were divided into several areas, in which the depth of heating in relation to irradiation time intervals was checked. The impact of individual heating phases on the distribution of the irradiated field and comparison of drying rate at different lengths of the cycle "heating-cooling". Theoretical potential of construction materials rehabilitation with the use of microwave radiation has been known for many years. This technique has been used by a number of companies in the construction practice. However, obtaining accurate information on suitable devices, procedures of their use, remedial action technique, positives and negatives and potential risks when using microwave radiation is very demanding since the findings obtained from practice are protected by individual companies for their own needs. It is understood that it is part of their "know-how". The long-term research based on close cooperation between Civil Engineering Faculty, BUT Brno, and S. P. UNI company, tries to make accessible enough information on potential and suitable use of the microwave technology in practice, based on laboratory and in-situ experiments.

1. Introduction

Microwaves is a term given to a part of electromagnetic irradiation characterized by wavelength ranging from 1 cm to 1 m. Microwaves spread into space from a source in the form of waves. Worldwide, a frequency of 2.45 GHz with a corresponding wavelength of 12.2 cm has been designated for industrial purposes [1, 2].

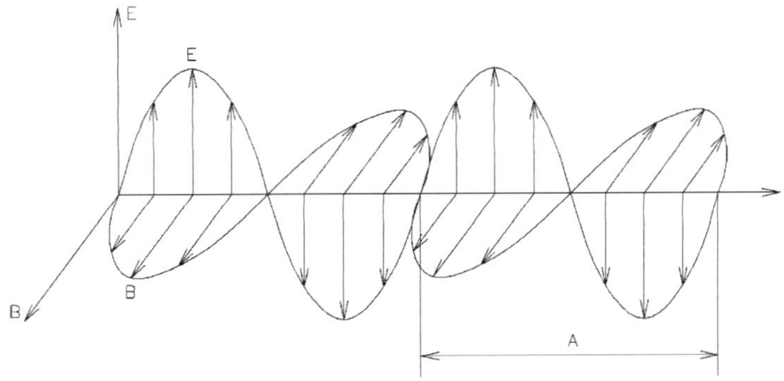

Figure 1. Electromagnetic wave with length λ – electric (E) and magnetic (B) components of the wave [3].

IOP Conf. Series: Materials Science and Engineering **728** (2020) 012007 doi:10.1088/1757-899X/728/1/012007

2. Drying by EMW irradiation

Despite having a bi-polar character, water molecules are electrically neutral. Therefore, in an electric field each and every water molecule is oriented by polarity (positive to negative and negative to positive), especially if the polarity alternates. This is precisely the case of microwave irradiation. The polarity of the electromagnetic field varies depending on frequency F, where the frequency can be higher than 10^9 times per second [1,2, 4].

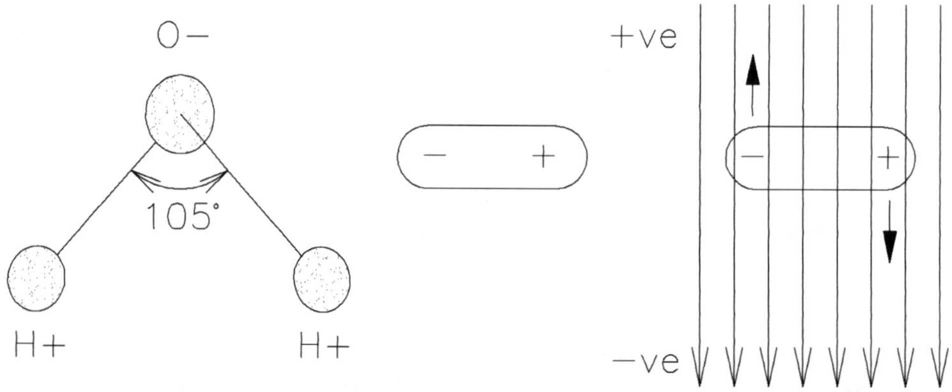

Figure 2. Water molecule and its orientation in electric field [6].

Heat is produced at high frequencies of radiation (i.e. oscillating vibration) and the subsequent change in orientation of molecules of loosely bound water. The thermal energy is produced by friction of molecules. This phenomenon is called polar rotation or polar friction [2]. In a simple manner it can be described as follows: the entry of microwaves into structures causes a rapid change in polarity of molecules of loosely bound water, causing their rapid movement in the material. Because of this movement particles themselves collide into each other, hence producing thermal energy. This process results in a conversion of water from liquid state into gaseous form, and its subsequent evaporation from the structure [2, 5].

3. Principle of drying variant

Two identical blocks with dimensions of 600 x 150 x 250 mm were used. The blocks were freely laid in a testing area of BUT Brno. After soaking the blocks were being dried. A different time interval of the heating-cooling cycle was selected for each block. However, total heating time was the same.

3.1. Course of experiment

Experiment procedure:

- weighing and marking of samples,
- weighing of samples with absorbed air humidity,
- immersion of blocks into a waterbath for 60 minutes,
- removal from water, 30-minute stabilization of absorbed moisture, packed in a foil for 22 hours,
- weighing of blocks after soaking,
- drying after time cycles,
- measurements of surface temperatures at the front and back side after heating cycles,
- another cycle of heating,
- measurements of surface temperatures after individual heating phases,
- weighing for 18 hours after the last heating phase [1, 2].

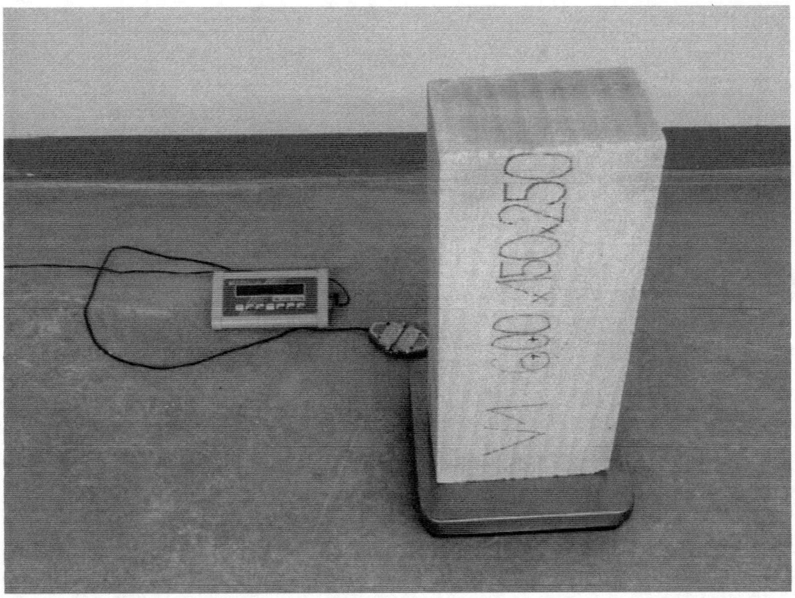

Figure 3. Weight of dry autoclaved aerated concrete sample V1, V2 before soaking.

3.2. Division of testing blocks

Block No. V1

- Soaked for 60 minutes in water, 60-minut stabilization of absorbed water, heating 4 x 15 minutes, a 15-minute break between heating cycles,

Block No. V2

- Soaked for 60 minutes in water, 6o-minut stabilization of absorbed water, heating 2 x 30 minutes, a 15-minute break between heating cycles.

Figure 4. Heating of samples V1, V2 by microwave generators MG electronic MB.

The values of surface temperatures measured in autoclaved aerated concrete blocks are shown in Table 1 and Table 2. They were identical cross sections of blocks, which were each exposed to different irradiation times (4 x 15 minutes, 2 x 30 minutes). Surface temperatures were measured at the front as well as at the back sides of blocks.

3.3. Partial experiment results

Table 1. Temperatures of samples V1 after heating intervals.

Block No. V1	Heating No. 1	Heating No. 2	Heating No. 3	Heating No. 4
Front side	80.9 °C	83.5 °C	91.9 °C	92.3 °C
Back side	32.1 °C	50.5 °C	59.4 °C	71.4 °C

Table 2. Temperatures of samples V2 after heating intervals.

Block No. V2	Heating No. 1	Heating No. 2
Front side	83.1 °C	88.8 °C
Back side	38.4 °C	61.1 °C

Table 3 shows decrease in weights of autoclaved aerated concrete blocks in "dry" condition, after soaking, and after 18-hour stabilization after final heating cycles. The weight decrease is almost identical for both blocks. The irradiation of these blocks revealed that shorter times of irradiation with a higher number of repeated cycles is much more effective.

Table 3. Weight decrease of autoclaved aerated concrete blocks after different heating intervals.

Block No.	Initial weight	Weight after soaking	Weight 18 hours after drying	Weight decrease
V1	12.510 g	14.015 g	12.717 g	1.298 g
V2	12.512 g	13.983 g	12.837 g	1.146 g

Figure 5 graphically displays initial weights, weights after soaking, after stabilization, and total decrease in weight of autoclaved aerated concrete blocks. It is obvious that the measured values of both autoclaved aerated concrete blocks are comparable.

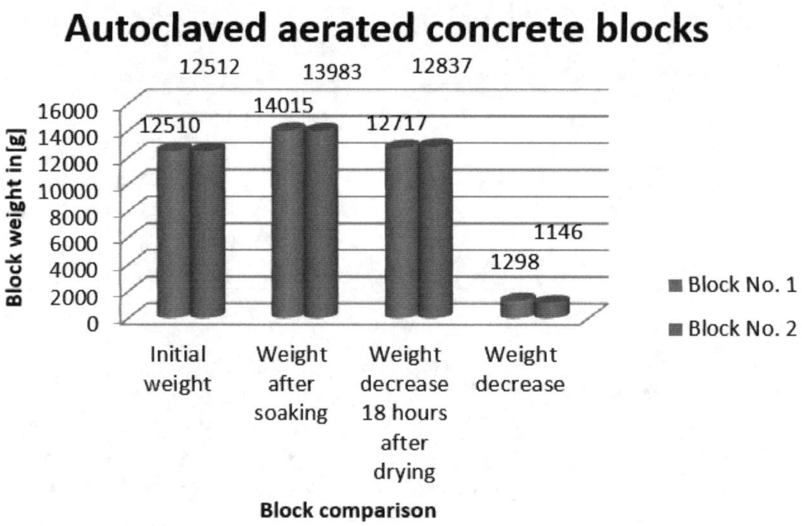

Figure 5. Illustrative block weight decreases after different heating intervals.

4. Experiment assessment

The experiment aimed to check the drying speed of autoclaved aerated concrete blocks in two different time cycles "heating-cooling". Total time of heating was the same for both blocks – 60 minutes. In the first case the cycles were divided into 4x15 minutes, in the other into 2x30 minutes. The measurement of surface temperatures at the front and back side, which was performed straight after the end of each heating proved that higher temperatures were reached with sample V1 (at the same time of active heating) than with sample V2. Regarding weight, higher decrease was found with sample V1; both in absolute values as well as in terms of percentage. The factor that affects this is the second phase of the cooling cycle. In this phase intensive moisture evaporation occurs from the surface layers, and thus conditions for deeper penetration of microwave radiation is created. This would apply for drying of all similar construction materials [6-8].

5. Conclusion

When irradiating, heating and drying all materials that were used within individual experiments, the frequently published opinion that the irradiated material is heated in depth from the very beginning was not confirmed. Moreover, it would contradict basic physical laws. The penetration depth of microwaves through the material depends on its moisture and distribution of moisture in the whole cross section. In cases when higher and evenly distributed moisture is concerned, microwave energy absorption occurs by surface layers first. Only afterwards, when moisture from these layers is reduced, the microwaves penetrate deeper. Therefore, during longer, uninterrupted, irradiation of a single spot, an undesirable gradient may occur between the front part temperature and temperatures in different cross section depths which may thus lead to uneven drying of individual layers. Consequently, undesirable stress of the heated and dried material or structure may occur, and a risk of shape deformations, or even destruction, may appear [3, 4].

Acknowledgment(s)

This paper has been produced within the project FAST-J-18-5506 - Study of drainage factors of retention roof compositions using recycled polyester boards.

References

[1] J. Sobotka, and R. Smolka "The temperature distribution during the microwave drying of wooden elements for structural defects in buildings," *MATEC Web of Conferences,* vol. 146, 03008, 2018.

[2] D. Bečkovský, L. Vacková, T. Bečkovská, J. Sobotka, J. Pěnčík, and M. Lavický "Analysis of the diffusional properties of Peripheral walls of wooden houses during Emw radiation exposure," *WOOD RESEARCH*, vol. 61, pp. 627-636, 2016.

[3] J. Plachý, "The analysis of implementation options of transverse joints of bitumen sheets," *MATEC Web of Conferences,* vol. 93, 02006, 2017.

[4] R. Smolka, and J. Sobotka, "Application of Recycled Plastic in Flat Roofs," *MATEC Web of Conferences*, vol. 146, 02009, 2018.

[5] M. Novotný, K. Šuhajda, J. Sobotka, J. Gintar, E.S. Dová, M. Mádl and Z. Jiroušek, "Use of microwave radiation in building industry through application of wood element drying." *WOOD RESEARCH,* vol. 59, issue 3, pp. 389-400, 2014.

[6] J. Sobotka, and R. Kolář, "Drying of the Basement Spaces of the Faculty of Arts in Brno," *Applied Mechanics and Materials*, vol. 861, pp.295-302, 2016.

[7] M. Procházka, and J. Sobotka, "Spreading of temperature field caused by microwave radiation," *Applied Mechanics and Materials,* vol. 824, pp.355-362, 2016.

[8] J. Sobotka, and R. Smolka, "Use of EMW radiation in the building industry at defects in buildings," *MATEC Web of Conferences*, vol. 93, 01008, 2017.

IOP Conf. Series: Materials Science and Engineering **728** (2020) 012008 doi:10.1088/1757-899X/728/1/012008

Degradation of PVC Waterproofing Membrane containing Plasticizers

Marek Novotný[1]

[1] Faculty of Architecture, Czech Technical University, Thákurova 9, 169 00 Prague 6, Czech Republic

marek.novotny.izolace@email.cz

Abstract. Durability and repairability are one of the basic conditions for the economically efficient use of building materials. The actual lifetime of PVC waterproofing membrane is not only affected by the ageing of the materials, but also by the ageing of the joints. Thus, a significant phenomenon in the evaluation of the lifetime of roof cladding is its realisation - welding. Plasticizers are used for the production of PVC waterproofing membrane. The migration of the plasticizers and the quantity of non-combustible residues (fillers) influence the service life (durability) and the possibility of subsequent repair. Methods used to determine the quantity of non-combustible residues and quantity of plasticizers are described in the paper. Samples were taken from roofs with degrading roofing layer. Based on the results, the lifetime of the PVC waterproofing membrane was determined in dependence on the plasticizers.

1. Introduction

The issue of ageing of the waterproofing membrane is currently very widespread. The most common materials of waterproofing membrane are bituminous waterproofing membrane and polymerized PVC waterproofing membrane (foils). The issue of material ageing and constructional details in asphalt strips were discussed [1-3]. Ageing of the material and constructional details of polymer foils are based on works [4-5].

In the framework of ageing of waterproofing materials [4], two basic phenomena have been identified which lead to premature destruction of the PVC waterproofing materials. There are plasticizers (plasticizers) and fillers (non-combustible residues). In the case of plasticizers, it is the migration (evaporation) of a substantial part of the plasticisers from the waterproofing membrane. In the case of fillers, this is an increase in the proportion, percentage of mineral fillers in the whole mass or on the surface of the foil materials.

The same destructive effect can be achieved by the correct ratio of the two components. In the case of synergy, very rapid destruction of the plasticized PVC waterproofing membrane occurs within a few years (See Figure 1). Materials of this type will fully survive the warranty provided by the manufacturer. If it is not possible to obtain a 10-year warranty from the manufacturer, it is clear that the materials will survive for a maximum of 5 years. Of course, these materials belong to the lowest price range, which is also the riskiest. The direct proportion applies, the lower the price, the shorter the service life.

For the reasons described above, the samples of plasticized PVC membrane were taken from the roof cladding. The samples were subjected to analysis to determine what was the dominant cause of the collapse of the softened PVC waterproofing membrane.

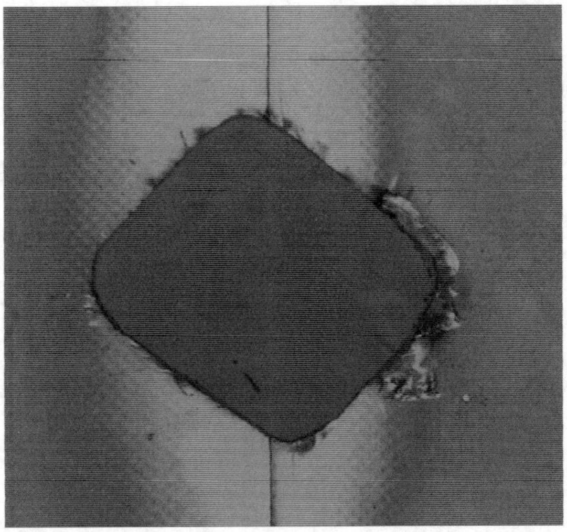

Figure 1. Patch applied to plasticized PVC waterproofing membrane that has already lost a significant amount of plasticizer [6].

2. Material and methodology

This paper is based on samples of the company A.W.A.L. [6]. Within the expert activity, many roofs with PVC waterproofing membrane were inspected and samples of polymeric membrane from softened (plasticizer) PVC were taken. These samples were subsequently analysed. Samples that had already spontaneous radial cracking were subject to inspection (See Figure 2).

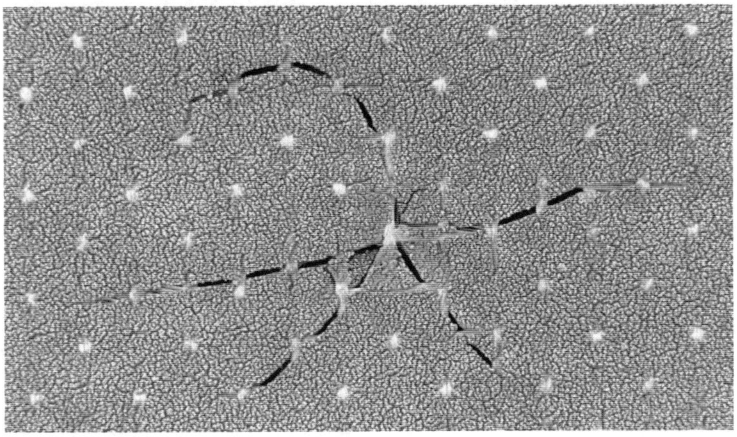

Figure 2. Exposed (uncovered) reinforcement insert on degraded softened PVC waterproofing membrane already showing the brittle fracture [7].

Basic chemical tests were used for all analyses. Specifically, these were IR (infrared analysis), gas chromatography, glass transition temperature analysis and a quantity of non-combustible residues. The paper is further specified by gas chromatography and determination of non-combustible residues. The analyses were carried out in cooperation with the Technical University of Liberec.

IR analysis shows changes in the entire waterproofing mass spectrum relative to any reference sample. The sample is obtained from the overlap of waterproofing, which is not exposed to UV radiation and other climatic stresses. Neglects the natural ageing, i.e. temperature, which affects the acceleration of ageing processes.

Gas chromatography provides quantitative results of a decrease in the volume of plasticizers in the mass of a softened PVC waterproofing membrane. This value is probably the most important and should be monitored. The decline of these values for the standard PVC waterproofing materials is observed.

The samples for analysis were taken from the supplied PVC waterproofing membranes (Figure 3). Each sample was carefully described. One gram of sample, and also a duplicate, was taken from the covered and uncovered part of every waterproofing membrane (together 4 samples of each membrane). Analytical balances Kern ABT 220-4M were used for weighing. Individual samples were cut into strips approximately 1 cm long and 3 mm wide and placed in 10 ml test tubes. Methanol (99.9%) was used as the extraction solvent. The volume of solvent added to the extracted sample was 10 ml. The samples prepared in the test tubes were placed on a horizontal back and forth shaker GFL 3006 and shaken for 70 hours. Subsequently, all samples were filtered using a glass syringe and 45 μm CHS Filterpure Syring Filtres filters. Because of the high sensitivity of the analyzing machine, samples were diluted 1000-fold before analysis.

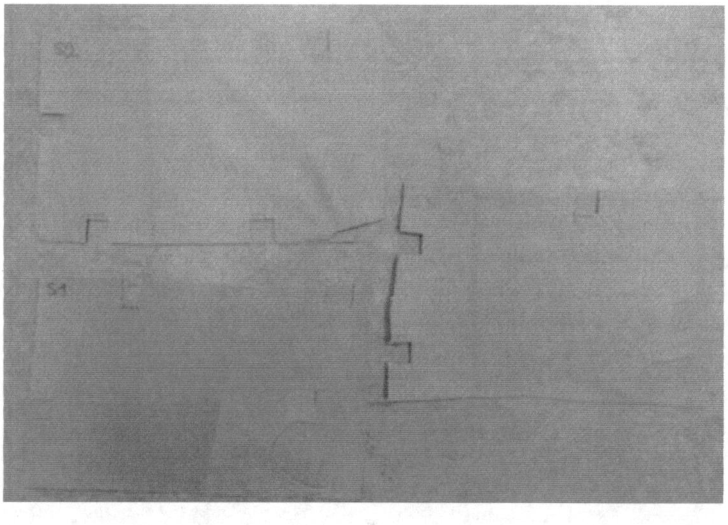

Figure 3. Analysed samples of waterproofing membrane from softened PVC [6].

Thermo Trace 1310 gas chromatograph with TSQ 8000 Triple quandrupole MS mass spectrometer was used to determine quantity of plasticizers. For separation, a chromatographic column DB-5MS was used with parameters: length 30 m, diameter 0.25 mm and a film thickness of 0.25 μm with a temperature limitation of 325°C/350°C. A liquid sample injection (1.0 μl) was injected by an autosampler PAL LHX-XT. The temperature mode of the gas chromatograph furnace was an initial temperature of 100°C, followed by a temperature gradient of 10°C per minute to 300°C, which has stabilized for 12 minutes. Minor oscillations of the MS system are removed by adding deuterated internal standard (ISTD - di-2-ethylhexyl phthalate D4) to all measured extracts.

The parameters of the measurement method were: ion source temperature 200°C, temperature transfer line 250°C, injection temperature 250°C and a carrier gas (helium) flow rate of 1 ml per minute. First, full-scale measurement of samples was performed, where all masses in the range of 35 to 500 m/z were scanned (m/z indicates the ratio between mass and charge and it is thus possible to identify individual substances). Based on the measured spectra and comparison with the NIST library, it was determined that probably the majority of the extracted component was diisooctyl and diisodecyl phthalate in all of the tested waterproofing membranes. Subsequently, the measurement method was modified with a focus only on the phthalate-specific mass 149, and the prepared extracts were analyzed and the mass 153, which is specific for the added internal standard.

Non-combustible residues, this test determines the quantity of the fillers in the tested material. A certain amount of mineral filler is in each film material, there they stabilize and improve technological and some technical properties, i.e. they are necessary. However, the amount depends on the entire composition of the waterproofing material. For plasticized PVC waterproofing membrane, the quantity of the fillers should not exceed 10%, which is a limit value where good quality waterproofing membrane can function properly in the long term [4-6]. The larger amount of fillers has a negative impact on the service life, but mainly on welding. The more filler in the surface layers, the worse the membranes are to weld. Above 10% this welding is already very risky [7].

The analytical balances Santorius BP110s were used to measure the sample weight. For each sample of welded foils, the content of non-combustible residues was monitored from two sampling - in the covered and uncovered parts of the bottom membrane material. The necessary parts of the membrane materials were separated from the tested waterproofing membrane. These were placed in pre-annealed ceramic crucibles (750°C), where the waterproofing membrane weights were in the range of 0.5 - 0.8 g. This weight corresponded approximately to a 20 x 20 mm area of the waterproofing membrane. The samples were then burned in an electric furnace (Nabertherm B180, Nabertherm, Germany) at 700°C/1.5 hours with air inlet. Ceramic crucibles, along with non-combustible residues, were left in the desiccator to cool freely to room temperature. The content of non-combustible residues was determined from the weights.

3. Results

The quantity of non-combustible residues is higher for the uncovered sides of the membranes than for the covered sides. The quantity of non-combustible residues on the covered sides was approximately 10.6%. The quantity of non-combustible residues on the exposed (uncovered) sides was approximately 12.1% (see Table 1).

Table 1. The quantity of non-combustible residues [6].

Sample	Sampling	Sample weight [g]	Weight of non-combustible residues [g]	Non-combustible residues [%]	Mean non-combustible residues [%]
S1 covered	1	0.7203	0.0769	10.68	10.75±0.140
	2	0.7119	0.0770	10.82	
S1 uncovered	1	0.5933	0.0702	11.83	11.94±0.223
	2	0.613	0.0739	12.06	
S2 covered	1	0.7659	0.0825	10.77	10.73±0.082
	2	0.7596	0.0812	10.69	
S2 uncovered	1	0.5968	0.0722	12.10	12.22±0.254
	2	0.5829	0.0720	12.35	
S3 covered	1	0.7581	0.0811	10.70	10.25±0.896
	2	0.7621	0.0747	9.80	
S3 uncovered	1	0.5853	0.0701	11.98	11.98±0.007
	2	0.5958	0.0714	11.98	

Calibration was not necessary for the relative comparison of the plasticizer (phthalate) content. The results are represented by the ratio between the measured area of the analytes on mass 149 and the area of the internal standard measured on mass 153, see Table 2. Table 3 shows the calculated percentage decreases in the content of diisooctyl phthalate and diisodecyl phthalate in the uncovered parts of the membrane versus the covered parts. The ratio of the area of the analytes to internal standard express the measured concentration. The higher the ratio, the greater the concentration in the sample.

Table 2. Calculated mean ratios of measured surfaces of analytes vs. ISTD [6].

Sample	The ratio of diisooctyl phthalate/ISTD	Measured area of diisodecyl phthalate/ISTD
S1- uncovered	6.7	35.9
S1-covered	11.2	90.7
S2- uncovered	6.3	43.4
S2-covered	12.2	116.9
S3- uncovered	7.0	60.6
S3-covered	12.7	118.6

Table 3. Calculated decrease [%] of plasticizers in the exposed (uncovered) parts of the membrane vs the covered parts [6].

Sample	Decrease of diisooctyl phthalate	Decrease of diisodecyl phthalate
S1	40.7	60.4
S2	48.0	62.8
S3	45.3	48.9

The presence of plasticisers was detected using the gas chromatograph Thermo Trace 1310 with TSQ 8000 Triple quandrupole MS mass spectrometer. Based on the measured spectra and comparison with the NIST library, it was determined that these were plasticizers diisooctyl phthalate and diisodecyl phthalate. Both plasticizers were included in all supplied samples of waterproofing membranes S1, S2 and S3. The decrease in the diisoocyl phthalate content ranged from 41 to 48% in the covered parts of the membrane compared to the uncovered parts of the membrane. The lowest decrease was observed in PVC waterproofing membrane S1 and the greatest decrease was observed for PVC waterproofing membrane S2. The loss of diisodecyl phthalate ranged from 49% to 63% in the covered parts versus the uncovered parts of the waterproofing membrane. The lowest decrease was found in membrane S3 and the largest decrease was found in membrane S2. The greatest decrease of both plasticizers was measured in PVC waterproofing membrane S2 (48 and 63%).

4. Discussion

Dependence of durability (functionality) of synthetic membranes on the loss of plasticizers can be determined in the evaluation of the statistics that are currently available. However, the condition is that the amount of fillers remains below 10%. In the case of a significant increase in the percentage of plasticizers, the durability will also be reduced below the limits stated here.

Figure 4 shows the results of dozens of laboratory tests of waterproofing membranes made of softened PVC. The tests were aimed at decreasing the percentage of plasticizers in the mass of the waterproofing membrane itself and the occurrence of cracks (spontaneous) or hail damage. Based on the analysis of the results, the risk of a decrease plasticizers between 20 - 30% (yellow rectangle) is assumed. In the case of a higher decrease of plasticizers, the cracks of the investigated waterproofing membrane materials have always been identified. The lifetime (durability) of a particular waterproofing membrane can be estimated depending on the decrease of quantity plasticizers. Red-coloured area shows safe behaviour of membranes during aging. The blue area shows the risk (very dangerous) behaviour of the membranes during aging.

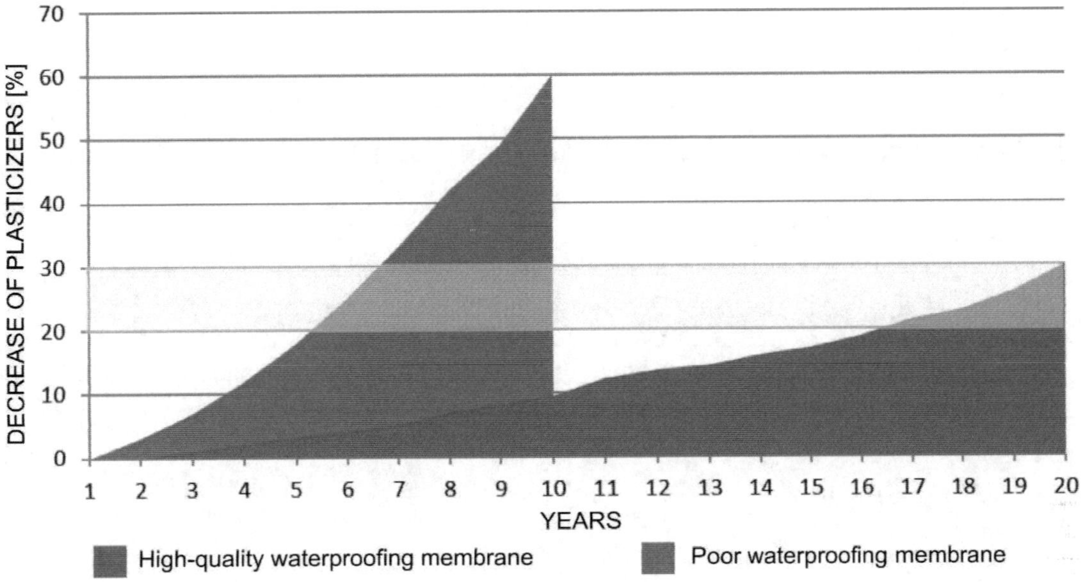

Figure 4. The lifetime of synthetic mPVC waterproofing membrane depending on the decrease of plasticizers [6].

5. Conclusions

The decrease quantity of plasticizers (compared to the overlapped material) can be observed in units of percent in the case of high-quality waterproofing membranes made of softened PVC, which will survive without problems for 20 - 25 years. On the other hand, in the case of defective waterproofing membrane materials, it is possible to observe a decrease of plasticizers by up to 60% in 10 years. This was always a collapse condition and the waterproofing material was plank and very brittle. Manipulating such material was impossible because it immediately cracked. The limit of quantity of fillers in softened PVC waterproofing membranes was confirmed. The quantity of the fillers should not exceed 10% if the service life is assumed. These underlying causes are the loss of sufficient technical properties to withstand the usual climatic stress and impossibility of repairs (i.e., the inability of the waterproofing membranes for further welding).

References

[1] R. Kunic, B. Orel, and A. Krainer, "Assessment of the Impact of Accelerated Aging on the Service Life of Bituminous Waterproofing Sheets," *Journal of Materials in Civil Engineering,* vol. 23, issue12, pp. 1746-1754, 2011.

[2] J. Plachy, "The problem of the compatibility of bitumen sheets for the reconstruction and rehabilitation of roofs," *MATEC Web of Conferences*, vol. 93, 02004, 2017.

[3] J. Plachy, J. Vysoka, and R. Vejmelka, "Insufficient dimensional stability of bitumen sheets as a source of flat roof defects," *MATEC Web of Conferences*, vol. 146, 02014, 2018.

[4] M. Novotny, and I. Misar, "Pathological changes waterproof membranes and their identification," *MATEC Web of Conferences*, vol. 93, 02008, 2017.

[5] I. Misar, and M. Novotny, "Defects and behaviour of inverted flat roof from the point of building physics," *MATEC Web of Conferences*, vol. 93, 02002, 2017.

[6] Company archive A.W.A.L., s.r.o.

[7] Personal archive of author Marek Novotny (Unpublished)

IOP Conf. Series: Materials Science and Engineering **728** (2020) 012009 doi:10.1088/1757-899X/728/1/012009

Mid-Rise Atrium Residential Houses - Domain of Comfort

Irina Maltseva[1], Natalia Kaganovich[1], Konstantin Tkachuk[2]

[1]Ural Federal University, Department of Civil Engineering and Architecture, Yekaterinburg, 620002, Russia

[2] Technical University of Munich, Institute of Energy Efficient and Sustainable Design and Building, Arcisstraße 21, 80333 Munich, Germany

3724316@mail.ru

Abstract. The paper discusses the role of a natural landscape in an industrial city, the harmonious interaction of an architectural object and the environment in creating a unified eco-architectural space. The authors focus on the typological aspect of residential architecture development, the feasibility of introducing low-rise complex development in the urban environment, suggesting the use of a morphologically autonomous category of urban villas as a type of residential building. The paper considers the advantages of a sustainable and presently relevant spatial archetype – an atrium – and its inclusion in the structure of a residential building and the eco-technical direction of its use along with separate engineering and technical solutions within the concept of energy saving. As an example, a low-rise residential complex project of an 'urban apartment villa' typological structure is considered, with the use of a multi-comfortable house system and the atrium principle of internal space organization in order to create an eco-friendly residential environment in the context of urban development and natural landscape.

1. Introduction

The general tendency for the development of urban areas and the growth of cities as a whole is accompanied by an increase in building density due to construction of high-rise buildings and concentration of urban infrastructure, including transportation systems, which leads to a deterioration of the ecological climate, living conditions and an increase in psychological stress in society. Within the development of the prestigious areas with new residential buildings, especially in the city centre, it is important to consider the ratio of the area and number of floors of buildings, the importance of landscaping and amount of vegetation in the area, the natural and urban context of the site, bioclimatic factors, object typology, and, of course, the possibility of applying modern "green building" technologies. This determined the goals and objectives of the study project: Residential complex "Zelenyj bereg" in Yekaterinburg.

2. Location

As a site, a triangular area was selected, bounded by Belinsky and Furmanov streets and the bank of the Iset River. The site is currently undeveloped and is intended for residential development. A complex of townhouses is located on the opposite side of the river, which determined the scale of nearby development. Municipal and industrial facilities located along the bank are gradually being

IOP Conf. Series: Materials Science and Engineering **728** (2020) 012009 doi:10.1088/1757-899X/728/1/012009

moved outside the city, freeing up the territory for residential high-rise buildings, emphasizing the guest route at the entrance to the city center.

2.1. River

Modern cities commonly began their history on the banks of rivers that served as transport routes and sources of water. Later, with the development of industry, the first factories and settlements grew along the rivers. The water area not only provides the city with a possibility for growth but is its center of attraction and a place of rest for inhabitants. The task of reviving and preserving the city's water landscape is also very relevant for Yekaterinburg: how to use the banks of the Iset river harmoniously and wisely, whether it is worth reserving the banks exclusively for recreation of citizens or use for construction, and how to link the scale of a small river to the scale of the surrounding buildings and the industrial city as a whole.

3. Typology

The choice of the type of building as a mono-space for the formation of a complex of buildings as part of an urban structure, is connected in this case with the urban planning (place) and natural (river) context, which determined the typological choice – a low-rise residential complex with a mixed urban structure "urban apartment villa" [1, 2]. Preserving all the advantages of individual housing and the continuity with the house-villa in its traditional sense, this type of residential building today has become an actual model of a modern residential and urban structure. This largely determines the very idea of an "ideal" home: first of all, the combination of comfort and privacy with the possibility of organizing public spaces, taking into account its small scale - the introduction of modern "green" technologies and energy-saving systems. The urban apartment villa has the economic advantages of a mid-rise apartment building: the building's wide-body, even with this number of floors, the compact organization of space, moreover, multi-factor loads on residential areas are reduced compared to high-rise buildings. The mixed structure of the building determines the variety of space-planning decisions of apartments, creates opportunities for the integration of landscape design into the border areas (landscaping of house sections, walls, roofs and terraces) and the interior spaces of a residential building.

3.1. Atrium

Atrium, as an alternative principle of the structural organization of space, has great functional and environmental potential [3, 4]. The atrium, as well in a residential building, is a multi-luminous integral space, mostly socially-communicational, that provides connection and relaxation of the inhabitants within the common area. As well it serves as a mean of increasing the compactness of the building and integrated residential development as a whole. With profound solutions and rational use of the atrium structure, the issues of building energy efficiency, creation of optimal microclimate and comfort of the internal environment are resolved. Among the aspects that shape the structure of modern buildings, the important one is creation of an effective ecosystem of the internal space, including the provision of natural ventilation and natural lighting, heating and cooling of rooms depending on the season and climatic conditions. In this regard, the atrium, as a historically stable model of the organization of the architectural space [5], as well as the general ecotechnical concept of the buildings, can significantly affect the level of comfort of the internal environment. In this case, two physical effects are considered: the greenhouse effect and the effect of dynamic traction [6]. It is the greenhouse effect that allows you to retain heat from sunlight and heat radiation from the inside in the atrium's space, which positively affects the microclimate in winter, but negatively in summer. Consequently, the use of the atrium features can be considered in two ways:

- in the winter, the internal air is heated due to the used air, and in the summer it is cooled due to natural ventilation;

IOP Conf. Series: Materials Science and Engineering **728** (2020) 012009 doi:10.1088/1757-899X/728/1/012009

- air exchange and natural ventilation are provided due to air traction. In general, this space is considered as an air reservoir and a "ventilation duct".

The principles of organizing natural ventilation in the atrium space are actively used in energy-efficient buildings, in an appropriate form, depending on the time of year and climate.

3.2. Comfort
In the twenty-first century, people will need a completely different standard of living. A "multi-comfort" house is the perfect home in terms of living comfort and maintaining an optimal indoor climate (comfortable temperature and humidity) with the "smart house" system. The environmental component in this context should be the creation of a multi-comfort environment in a "passive" house.

3.3. Energy efficiency
The energy efficiency of the house in the building construction and operation practice aims to reduce the level of energy consumption throughout the entire life cycle of the building. "Passive" is a house in which the energy consumption for heating is minimized (down to zero) with the help of internal heat sources, modern energy-saving technologies and highly efficient heat-insulating materials [7-9]. In such a building, all energy processes should be optimized. From modern standards, truly energy-efficient houses are the ones that have enough energy from individual renewable sources for hot water supply and heating. The specific heat energy consumption for heating a passive house, determined using the "Passive House Planning Package" (PHPP), should not exceed 15 kW•h/(m^2•year). The total consumption of primary energy for all domestic needs (including heating, hot water and electric energy) should not exceed 120 kW•h/(m^2•year). Moreover, the passive house should be comfortable in the summer without the use of an air conditioner!

In the case of designing a low-cost passive house, the main points should be considered: improvement the thermal insulation of standard building elements (roof, walls, floors), reduction of thermal bridges, sealing of building's shell, usage of special windows for passive buildings, efficient heat recovery from exhaust air.

4. Design decisions
As a result of the analysis of the selected situation, taking into account a number of factors, a general project concept was developed, consisting of three main areas.

4.1. Urban planning concept
Urban planning concept:

- improvement of the river bank – the creation of a promenade with the competent use of the natural landscape;

- residential development – urban apartment villas as a transition from the scale of a low-rise residential complex on the opposite bank of the river to a multi-story development of adjacent territories;

- compositional development solution – includes landscaping of the residential complex and terraced construction of the coastal promenade, taking into account its natural context (scale and bending of the river, shoreline geometry, existing terrain) and providing clear functional zoning of the site and autonomy, while maintaining convenient interconnections;

- privacy and comfort – location of urban villas away from noisy highways on the upper level of the coastal relief, "protected" by high-rise buildings along the front of adjacent streets, preserves the privacy of the home, the psychological and environmental attractiveness of the site, and, taking into account the low density of the building, creates comfortable living conditions.

4.2. Spatial decisions of a residential building

The idea behind a compositional and space-planning solution of a residential building is the desire to harmoniously combine urban planning and natural surroundings with the typology and principle of organizing living space. The urban apartment villa with the combined planning structure and an internal atrium has a complex shape: a sculptural form and the "opening" of the volume towards the river and the straight geometry of the volume from the urban area. The planning structure is conditionally divided into three "radial" functional zones: a central atrium hall with a minimum external fencing area (ideal for thermal protection, an atrium model) [10] and upper glazing; buffer zones of utility rooms located around the hall and residential apartments zone (figure 1). Apartments with the number of living rooms from 1 to 5 have spacious loggias and balconies with container gardening, which compensates for the advantages of an individual dwelling. Ground floor apartments have own entrance from landscaped terraces. Hall-atrium with floor galleries, open green roofs, various public premises on the 1st floor, including a billiard room, mini-kindergarten and fitness center form a single recreational and communication public space of the "club" house, which largely shapes the nature of the communication of its inhabitants. The constructive solution of the building provides the flexibility of the internal space and the possibility of individual planning decisions in each apartment and redevelopment options during inhabitation. To save territory, ensure environmental safety and increase the comfort of living, a unified underground automated parking for personal vehicles is provided in the complex.

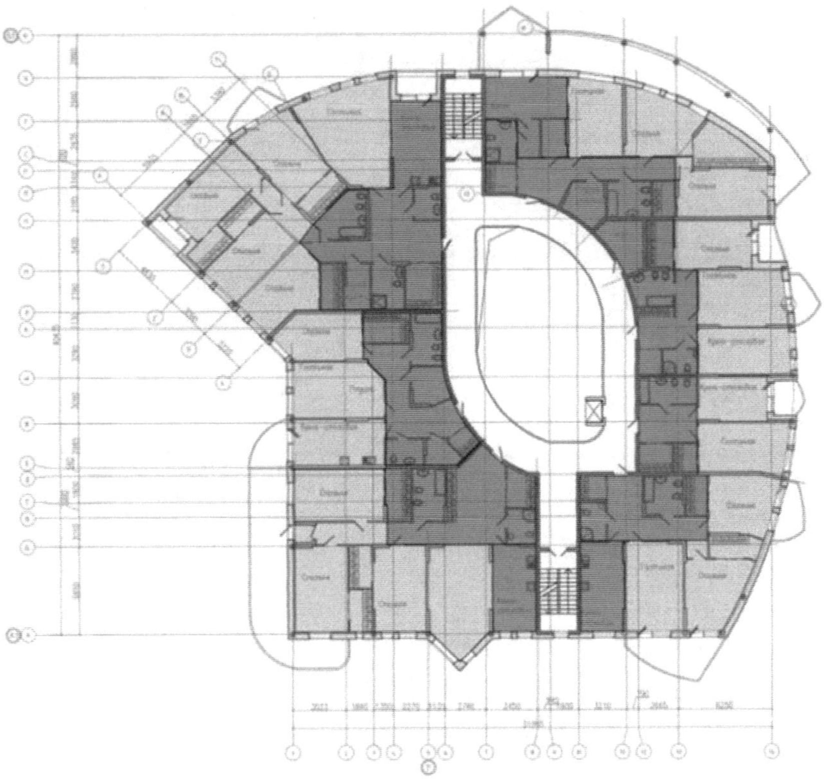

Figure 1. Floor plan: yellow - living rooms; dark green - bathrooms and corridors.

4.3. Energy-saving concept

The project adopted modern energy-efficient technologies. Solar collectors located on the operated roof contribute to the production of thermal energy, the excess of which is accumulated and stored in seasonal and daily heat accumulators. “Solar trees” are a source of electric energy, they are formed from arrays of innovative rotating solar batteries (cones with a power of 1000 W). They harmoniously fit into the environment and the energy supply from one tree is 9383 kWh per year. The project also proposes the use of a ground heat exchanger.

The elimination of cold bridges in monolithic concrete at the loggias joint was made based on Schoeck solutions, which reduces heat loss, prevents condensation and the development of the mold.

The windows of passive buildings work like solar collectors - heat from the passive use of solar energy makes the main contribution to the compensation of heat loss. Roller blinds with an automated control system SomfyAnimeoIB + allow optimal control of heat input and loss during interaction with the ventilation system. According to the principles of designing energy-efficient houses, the thickness of thermal insulation with a heat conductivity coefficient $\lambda = 0.04$ W/(m·C) should aim for 40 cm (min 25 cm), and the heat transfer resistance should be 10. The project adopted a thickness of insulation in the outer walls of 40 cm, in the coating (more efficient insulation) – 30 cm. For comparison: the required thickness of the thermal insulation of a residential building, as calculated in the Urals, is 170 mm. Of course, this is a large thickness of insulation, but only by following this path, it is now possible to achieve small heat losses at an affordable price.

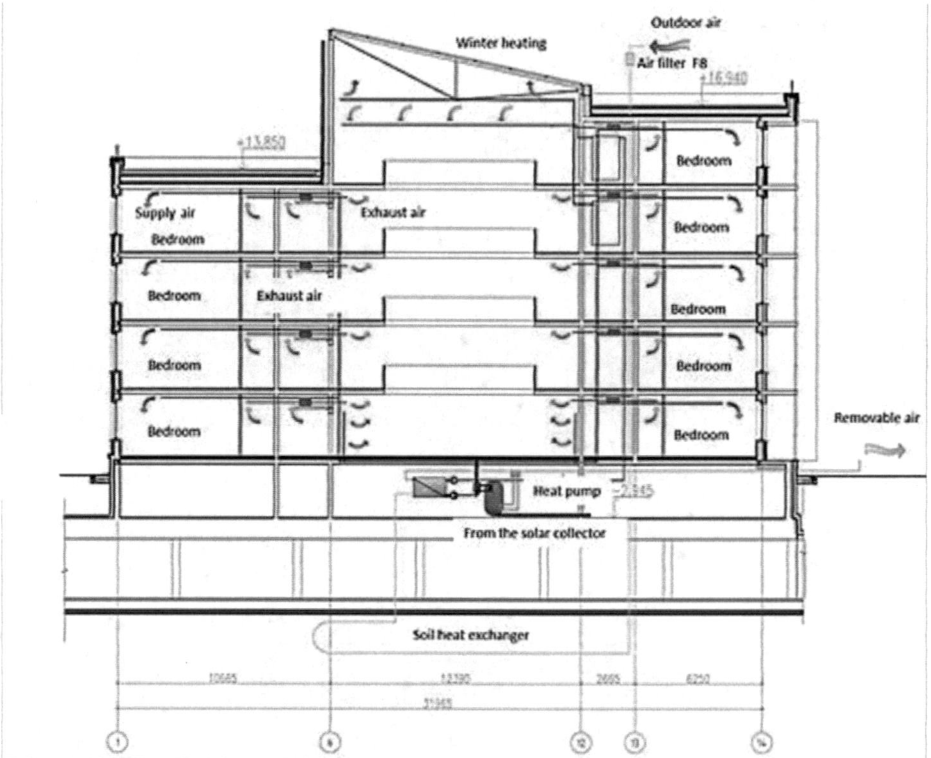

Figure 2: Ventilation scheme.

The ventilation system (figure 2) is divided into three zones: the first is the air supply zone for residential premises, the second is the transitional zone around the perimeter of the utility rooms, hallways, corridors, stairwells located around the atrium, the third zone is the exhaust air zone where all wet rooms are connected. The used air from domestic premises is sent to the atrium, due to which additional heating of the air in it is carried out. The second important condition remains that the ventilation unit, if certain requirements for its maintenance are followed, serves primarily for air hygiene. Also, the use of recuperators will provide our building with a high share of heat recovery (more than 75%), which is a prerequisite for passive buildings

5. Conclusions

The following engineering and technical decisions were made in the project:

- modern environmentally friendly building materials and structures are used to create an environmentally sustainable living environment;
- independent energy sources are used for the effective functioning of all engineering systems of a residential building;
- continuous insulation building envelope principle is applied;
- installation of green roofs and container gardening of terraces and loggias using latest technologies in this area.;
- residential development.

Today, housing in Yekaterinburg impresses with its scale and pace of development. It is important that our houses are not only comfortable, beautiful, energy-efficient but also that they organically fit into urban development and the natural landscape since it is the architect who must teach the inhabitants the language of landscape and communication with nature.

References

[1] A. V. Kalabin, "Typology of Residential Buildings, Small and Medium Number of Storeys: Current Status," *In Rus: RAASN J., Academic Bulletin*, Vol. 1, pp. 63–69, 2014.
[2] D. V. Berezin, A. V. Merenkov, "Apartment villa: a new type of urban housing," *In Rus: Chelyabinsk: SUSU,* 104 p., 2011.
[3] Environmental factors in the design of public buildings. Environmental design as a modern way of forming the volume planning structure of the building, *In Rus.:* https://studref.com/303291/stroitelstvo/sredovye_faktory_proektirovanii_obschestvennyh_zdaniy_ekologicheskoe_proektirovanie_sovremennyy_put_fo,
[4] E. M. Khorovetskaya, "Atriums in Modern Architecture," *In Rus: 13 Int. Scien. Conf. "Seifullin Readings–13": collection of materials*, vol. 1(5), pp. 178–181, 2017.
[5] E. B. Ovsyannikova, "Architectural typology," *In Rus: Yekaterinburg: Tatlin,* pp. 14–17, 2015.
[6] R. Saxon, "Richard Atrium buildings," *In Rus: Moscow: Stroyizdat,* pp. 98–107, 1987.
[7] V. Feist, "Basic provisions for the design of passive houses" (Translation from German with additions edited by A.E. Elokhov), *In Rus: Moscow: ASV,* pp. 67-91, 2015.
[8] Schöck Design Guide: Solutions to Prevent Thermal Bridging, https://www.schock-na.com/view/5752, 2014
[9] R. E. Perminov, I. N. Maltceva, N. N. Kaganovich, A. Yelohov, "Energy-efficient multi-comfort group of buildings in Svetlorechenskiy estate," *J. Appl. Mech. and Mat.*, Vol. 725-726, pp. 1389–1394, 2015.
[10] D. Zhukov, "Atrium houses as an example of eco-architecture," *In Rus: Internet J. Construction expert*, https://ardexpert.ru/article/22, 2015.

Smart Green Bridge - Wildlife Crossing Bridges of New Generation

Jaroslav Žák [1], Michal Kraus [1], Petra Machová [1], Jan Plachý [1]

[1] Institute of Technology and Business in České Budějovice, Department of Civil Engineering, Okružní 517/10, 370 01 České Budějovice, Czech Republic

info@krausmichal.cz

Abstract. The research content of the paper is focused on landscape-vegetation and technical-economic optimization of vegetation elements of buildings. The length of the motorway in the Czech Republic has increased approximately three times since 1995. Road transport infrastructure is constantly growing. Ensuring the passage of important transport structures for fauna is an objective requirement. The function of migration corridors is ensured by wildlife crossing bridge. These constructions are currently based on concrete and steel, which is not only environmentally friendly but also presents several other problems. The paper presents the idea of a new smart solution for wildlife crossing bridge. Smart Green Bridge is based on principles and technologies following the sustainable development such as the use of load-bearing timber structures, extensive greenery of about 100 mm thickness, vertical greenery, self-guiding vertical and horizontal irrigation systems, photovoltaic panels, battery energy storage and next.

1. Introduction

Urban areas show significantly higher temperatures than other areas located in the undeveloped surrounding area in the context of climate assessment. This is due to the overlapping of large areas of the original vegetation with buildings and roads. Concrete and asphalt surfaces absorb a huge amount of heat from the Sun, which then radiates back into the environment. The overheating of cities causes the rise of warm air, which raises dust particles that people inhale. Due to the ever-expanding urban development, the intensity of the so-called thermal islands is increasing and life in some cities becomes even more difficult in the summer months. For example, the thermal islands in the center of Prague show a temperature of more than 2°C higher than the temperature in the outskirts of the city. Due to the absence of green areas in cities, natural run-off points disappear and water is very difficult to retain on concrete or asphalt surfaces and with its help to cool the surroundings. One possible solution to prevent this problem is to design more green spaces in cities. The green facades or green roofs can retain water and subsequently evaporate it. However, it must be emphasized that greenery on buildings also benefits the building itself. Adaptation measures in cities aim to adapt to climate change. This can be achieved by sustainable urban development while maintaining the necessary quality of life for the population. Water management also plays a very important role. It is necessary to focus on water retention and slowing down of water runoff from hard surfaces. It is also important to ensure the seepage or use of rainwater and to functionally connect the areas with the prevailing natural components. Water and vegetation areas with elements that significantly affect the

IOP Conf. Series: Materials Science and Engineering **728** (2020) 012010 doi:10.1088/1757-899X/728/1/012010

microclimate will play an important role. Their basic mechanism is water evaporation, which has the effect of lowering the ambient temperature.

Construction, and in particular the operation of traffic linear constructions (roads, motorways, railways), means hard negative impacts on the landscape and related ecosystems. The effects of linear transport construction on wildlife populations have been the focus of many studies in the last years [1-5]. The violent fragmentation of the landscape environment, connected with migration retardations and ecosystem vegetation barriers, is essential. The usual compensatory measures are technical objects, so-called wildlife crossing bridges or ecoducts. These are under-ground passages and over-ground crossings for the zoo-component and synergistic transfer of vegetation. The relevant buildings are based on migration corridors. Heavy reinforced concrete constructions are typical, especially above traffic road constructions. These constructions reflect thousands of tons of concrete, other technical elements, and great construction, energy and financial demands.

2. Wildlife crossing bridges - ecoducts

Nowadays, with the ever-expanding road network, not only is there a reduction in intra-urban traffic, but many of negative environmental impacts arise. Wildlife crossing bridges are now widely used they are abundantly built in many countries around the world [6]. Road traffic systems were developed to allow relatively easy, fast and effective transport of people and goods. However, it is precisely the impact of densifying transport infrastructure and moving people to larger cities that results in environmental restrictions, especially for animals. The consequence of these factors is the disintegration of the landscape. Table 1 shows the development of the road transport infrastructure in the Czech Republic from 1995 to 2018.

Table 1. Road transport infrastructure in the Czech Republic from 1995 to 2018.

Year	Motorways [km]	Expressways [km]	Roads [km]	Total [km]
1995	414		55 086	55 500
1996	423		55 088	55 511
1997	486		54 908	55 394
1998	499		54 895	55 394
1999	499		54 933	55 432
2000	499	299	54 909	55 408
2001	517	300	54 910	55 427
2002	518	305	54 904	55 422
2003	518	320	54 929	55 447
2004	546	336	54 953	55 500
2005	564	322	54 945	55 510
2006	633	331	54 952	55 585
2007	657	354	54 927	55 584
2008	691	360	54 963	55 654
2009	729	370	54 990	55 719
2011	734	422	55 018	55 752
2012	745	427	54 997	55 742
2013	751	442	54 965	55 716
2014	776	458	54 985	55 761
2015	776	459	54 972	55 748
2016	776	459	54 962	55 738
2017	1 223	0	54 535	55 757
2018	1 240	0	54 517	55 756

IOP Conf. Series: Materials Science and Engineering **728** (2020) 012010 doi:10.1088/1757-899X/728/1/012010

Transport infrastructure in the form of motorways and expressways is a big and insurmountable barrier for animals. This phenomenon is negative and undesirable because the landscape that is still working is then divided into smaller and even smaller parts. The result of this phenomenon would be the construction of migratory objects in the form of ecoducts. Wildlife crossing bridges would be placed in problem areas. Wildlife crossing could be divided into two basic group - overpasses and underpasses [7].

Figure 1. Wildlife crossing bridge, Czech Republic, Dolní Újezd, Motorway D35, realization 1999.

Currently, there are the following possible construction solutions for wildlife crossing bridges:

- Monolithic beam construction with the top plate. The basis is a load-bearing structure consisting of a combination of a monolithic beam (one or more) and a monolithic slab. Depending on the span length, the structure is designed as reinforced concrete or prestressed.
- Beam structures are based on prefabricated prestressed beams coupled with a monolithic reinforced concrete slab.
- The second type of beam structures consists of steel beams coupled with a monolithic reinforced concrete slab. It differs from the previous construction in that steel beams are used instead of prefabricated prestressed beams.
- Reinforced concrete monolithic structure in the form of a vault.
- Monolithic arch construction. In order to ensure the transfer of the arc force from the horizontal component, a prestressed plate in the form of a tie rod is designed for the structure (the prestressed plate is placed above the arcs). In order to ensure the transfer of forces to the drawbar, the structure is supplemented by struts in the area at the edge supports of the arches. The whole solution of the migration object involves a relatively complicated structural arrangement, the overall static functioning depends on the exact functioning of all the partial components forming the overall structure.
- Prefabricated reinforced concrete arch construction. The wildlife crossing bridges is based on a supporting structure consisting of a precast reinforced concrete structure in the shape of a vault. This construction is made up of the prefabricated wall and arched parts which are supported by an articulated joint. In order to ensure the interconnection of all components, a monolithic casting of the realized joints is made.
- Construction of the composite arch. The structure consists of arched beams, which are made of glued wood.
- Steel flexible construction. The overall supporting structure of the building is made of a flexible steel structure made of corrugated sheet metal, together with a vault made of compacted gravel sand [8].

Disadvantages and weaknesses of current structural and construction solutions of wildlife crossing bridges include:

- The major parameter of the negative biotic effect of the motorway under the wildlife crossing bridge is the wide width of current crossing structures.
- It is not possible to use standardized construction systems for the construction of current wildlife crossing bridges because these are individual projects.
- In order to ensure the growth of vegetation located on the surface of the wildlife crossing bridge, it is essential that the layer of soil is formed in a high thickness (mainly if it is a vaulted structure).
- Heavy and massive structures are the result of high loads.
- The retrofitting of existing wildlife crossing bridge constructions cannot be achieved without the necessary restrictions on the traffic on the crossing road [8].

3. Smart Green Bridge

The basic strategy of access and innovation of the above-ground ecological transition, the wildlife crossing bridges, is a light bridge structure with marginal walls created by vertically cultivated vegetation, separating transferring individuals and communities from the operation of the crossed barrier road. For the previously applied solutions of aboveground concrete crossings, their vegetation effect required at least 1 meter of the soil layer, which allowed cultivation, but only lower forms of vegetation (shrub layer). Recent soil layers, due to directional runoff conditions, are not aqueous, vegetation suffers from drought and gradually disappears. The construction is a completely foreign element for the landscape.

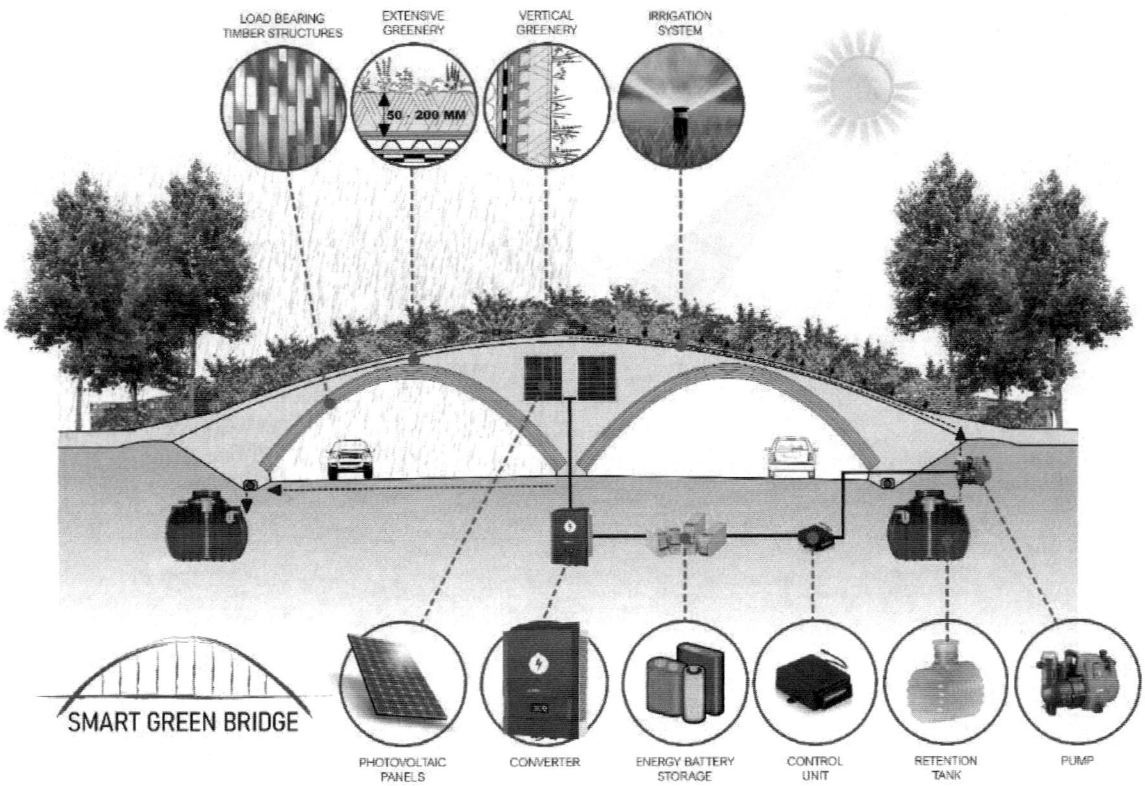

Figure 2. A pilot scheme of Smart Green Bridge.

The new solution is illustrated in Figure 2. The new solution of the so-called light bridge construction is sufficient with a surface thickness of approx. 100 mm with a purposefully created water regime based on renewable energy sources. This solution means great relieving of the wildlife crossing bridges, drastically reduced material and energy demands, greater flexibility and economic efficiency. The secondary effect is the use of vertical - high greenery in the immediate vicinity of the wildlife crossing bridges, creating quiet places utilizing the use of the construction and thus its landscape and social meaningfulness. Efficient use of rainwater from the bridged road to ensure irrigation of vegetation is essential. Collecting water from the line structure (road or highway) will be collected into the retention tank and distributed to the horizontal and vertical vegetation using the soil moisture sensors and by own energy source (solar photovoltaic panels).

The design of a new smart green crossing bridge solution means low demands on the technical solution, higher financial efficiency of the construction and minimal intervention in the construction and construction of the bridged road. From the landscape point of view, it is also possible to minimize the risks of anthropic impacts in the perimeter of the crossing bridge, and, for example, to increase the frequency of realized crossings on line constructions for economical sums.

The proposed structures, in terms of civil engineering, will be considerably lighter and simpler, standardized prefabricated structural elements and routine proven construction methods can be used. Technically and technologically, the procedure of their additive placement on already realized line constructions will be elaborated without any significant limitation of the existing operation. High material, energy and financial efficiency, the variability of utilization and positive ecosystem and landscape-creative impacts are predicted and documented. Figure 3 illustrates the visualisation of Smart Green Bridge.

Figure 3. Visualisation of Smart Green Bridge.

4. Conclusions

Concerning the financial and technological demands of existing wildlife crossing bridge, it is necessary to demonstrate objectively the necessity of a wildlife crossing bridge under the given conditions, but also to examine all variants of the technical solution with an emphasis on the cost-optimal level and principles of sustainable development. The existing construction methods are very inefficient. Criticism of their construction is often justified. This contribution presents the new generation of wildlife crossing bridge, the so-called Smart Green Bridge. Smart Green Bridge is based on principles and technologies following the sustainable development such as the use of load-bearing timber structures, extensive greenery of about 100 mm thickness, vertical greenery, self-guiding vertical and horizontal irrigation systems, photovoltaic panels, battery energy storage and next.

Acknowledgment(s)

The contribution is supported by the Specific University Research SVV 201907 Technical-economic optimization of vegetation elements of buildings.

References

[1] S. Pell, and D. Jones, "Are wildlife overpasses of conservation value for birds? A study in Australian sub-tropical forest, with wider implications," *Biological Conversation*, vol. 184, pp. 300-309, 2015.

[2] M.P.O. Olsson, P. Widén, and J.L. Larkin, "Effectiveness of a highway overpass to promote landscape connectivity and movement of moose and roe deer in Sweden," *Landscape and Urban Planning*, vol. 85, issue 2, pp. 133-139, 2008.

[3] A. Benten, T. Hothorm, T. Vor, and Ch. Ammer, "Wildlife warning reflectors do not mitigate wildlife–vehicle collisions on roads," *Accident Analysis & Prevention*, vol. 120, pp. 64-73. 2018.

[4] M.P. Juijser, E.R. Fairbank, W. Camel-Means, J. Graham, V. Watson, P. Basting, D. Besker, "Effectiveness of short sections of wildlife fencing and crossing structures along highways in reducing wildlife–vehicle collisions and providing safe crossing opportunities for large mammals," *Biological Conservation*, vol. 197, pp. 61-68, 2016.

[5] D. Beben, "Crossings Construction as a Method of Animal Conservation," *Transportation Research Procedia*, vol. 14, pp. 474-483.

[6] L. Corlatti, K. Hackländer, and F. Frey-Roos, "Ability of wildlife overpasses to provide connectivity and prevent genetic isolation," *Conserv. Biol*, vol. 23, pp. 548-556, 2009.

[7] P.A. White, "*Getting up to speed: a conservationist's guide to wildlife and highways,*" *Defenders of Wildlife*, 2007.

[8] R. Bocek, H. Kurth, and V. Křístek, The new construction system of ecological bridge (In Czech. *Nový konstrukční systém ekologických mostů*).

IOP Conf. Series: Materials Science and Engineering **728** (2020) 012011 doi:10.1088/1757-899X/728/1/012011

Rehabilitation and Reinforcement of Ferroconcrete Walls with Reinforced Pilasters

Michal Novotný [1], Jan Všetečka [1]

[1] Vysoké učení technické v Brně, Fakulta stavební, Ústav technologie, mechanizace a řízení staveb, Veveří 331/95, Brno, 60200, Czech republic

novotny.m1@fce.vutbr.cz

Abstract. The article is focused on renovation, rehabilitation and reinforcement of high ferroconcrete walls. It tries to point out the experience of experts in solving problems with high ferroconcrete walls. During a revitalization of buildings of a military accommodation facility in Brno with basement and connecting corridors of an average height of 25 m, degradation of concrete, locally missing reinforcement and poor-quality concrete mixtures, the strength class of which varied between C8/10–C10/12, were found in these corridors. Along the height of the pilasters, the structure also showed the signs of torsion and it was assessed the structure also needs to be strengthened/reinforced besides the rehabilitation. The reinforcement structure had to be built from the very foundation of the building, it is linked to the existing wall and has three cascade levels. For the first two parts, a construction pit has been created with a Berliner wall, and these parts reach below the modified terrain, are reinforced with concrete reinforcement according to the structural project, concreted with cast concrete, and the foundation is reinforced with deep-based construction. The second and the third parts of the cascade stretch from the modified terrain between the pilasters to the top and they are made with a two-stage shotcrete technique, the first one has inbuilt concrete reinforcement with welded wire mesh and the second one composite mesh anchored by helical reinforcement without static effect. In terms of the damage of the structure, carbon fibre wrapping technique or grouting of the existing ferroconcrete structure was also used in the affected areas. The article also includes photo documentation of the repairs and structural damage prior to them.

1. Introduction

The article was based on construction changes to the buildings of the Military Accommodation Facility in Brno. The project addresses the rehabilitation of the supporting structures of the connecting corridors in the building of VUZ Brno in Chodská street between the halls of residence buildings A1–A4 and B1–B2. The existing complex of six buildings has 6 to 9 floors (1B + 6–9F), the buildings are interconnected by a corridor on all floors. Interconnection of individual buildings is solved by a straight staircase or a lift shaft, because the neighbouring buildings are offset by one half of the floor height.

2. Structural system of the buildings

The structure of the connecting corridors, as well as the whole building, was designed as a monolithic ferroconcrete structure. Among other things, the corridors also function as a longitudinal reinforcement for adjacent buildings built as a framework system. Individual buildings are designed in a modular grid of 6.0 m in both transverse and longitudinal directions. They have three modules in the transverse direction and six modules in the longitudinal direction. Prefabricated ceiling panels are laid on cross girders. The foundation of the building is done via transverse and longitudinal girders in combination with the base slab. It is a variant of interconnection of the base slab and the grid foundation made up of wall footing. The connecting corridors that are currently rehabilitated consist of wall columns, or pilasters, connecting a monolithic wall. The vertical walls between the pilasters have a thickness of 0.4 m and a width of 3.8 m. The cross-section of the columns is 0.5 x 0.7 m. The perimeter walls and columns of the corridors are reinforced on each floor of the adjacent building by transverse girders of 0.5 x 0.7 m.

Figure 1. Damaged structure of connecting bridge with detail of reinforced concrete structure (damaged) [1].

3. Failure description

In the case of structures, structural damage is evident at first sight. The visual aspect shows high degradation of the structure's surfaces. The defects and failures are mainly formed by fallen concrete covering layer of the reinforcement and the associated and increased corrosion of the main bearing reinforcement and stirrups, which are often even non-functional due to the damage. In many places, the surface layer has not yet fallen, but it is evident from the cracks that moisture and the environment reach the very reinforcement. During the construction and technical survey, samples (core drilling, hacking off, collection of fallen parts) were taken for laboratory tests. It was also possible to apply non-destructive methods to analyze the state of the structure - for example, georadar or ultrasound-based detectors.

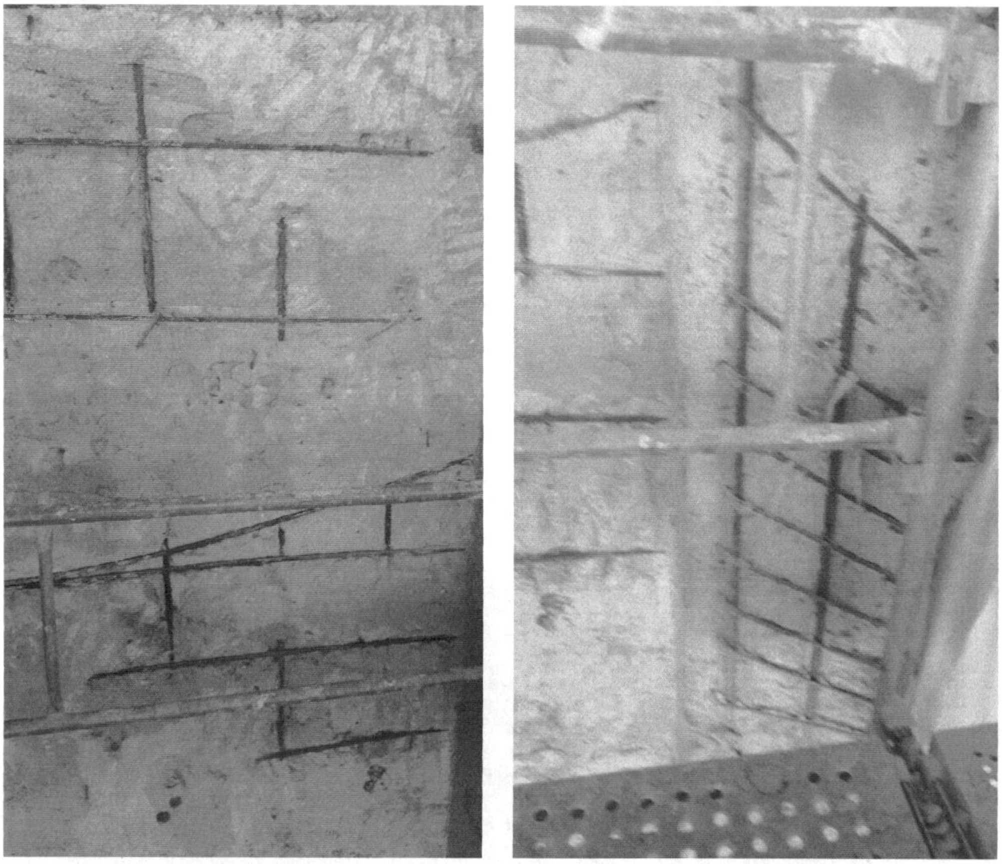

Figure 2. Exposed reinforcement imperfections in the structure - various incorrect positions, missing reinforcement, reinforcement degradation.

The complete list of used instruments and methods is not available to the author of the article, it is more or less the "secret" of the author of the exploratory and static part of the documentation Ing. Jan Perla. However, it is possible to find a description of the various methods either in the literature or scholarly articles, for example here - in the article "Identification of the position of concrete reinforcement" [2, 3].

The results corresponded to the observed state, the cylindrical strength of the concrete mixture varied from 6.7–17.5 MPa, being 8.4–21.9 MPa after conversion to cubic strength. Existing concrete therefore corresponds to concrete of class C10/12. Further, after mechanical cleaning and blasting of the structure, more problems with the structure were discovered. First, locally missing reinforcement in the form of bearing horizontal reinforcement in the wall area between the pilasters was found. Subsequently, missing or very unevenly/poorly placed stirrups were also found in the pilaster structure.

These places – both the missing bearing reinforcement and the stirrups – were often at the points of the dummy joint, i.e. at the point of reinforcement contact, where the transfer of forces from one reinforcement to another must work properly. Another found defect, which contributes to the structure degradation, was the technological indiscipline in concreting and laying of concrete, which resulted in separation of the concrete additive and binder, where only a ballast bed was often found above the dummy joint, with a small amount or completely missing binder. The condition of the structure is very poor and due to the long-term durability, safety and functionality of the structure, it is necessary to properly rehabilitate and reinforce the structure.

4. Proposed measure – repairs and reinforcement of reinforcing walls

The method of securing individual connecting corridors will be very similar – only the missing reinforcement which will be subsequently replaced locally by carbon wrapping with reinforced metal flanges and its extent/amount are assessed, and the technological indiscipline in the form of ballast beds is repaired by grouting. For each wall, a construction pit has first been done due to the depth of the foundation and the building having basement, where the securing must have been connected to the existing foundations. In the construction pit, the existing foundation structure has been reinforced and the deep foundation for the newly formed reinforcement of the connecting corridor structure done by performing piling work with inserted rolled profiles. The last structural part of the new reinforcement system will be the reinforcement of the front walls between the pilasters in the form of reinforcement and meshes [4-6].

5. Proposed repair work

Depending on the extent of the damage and its location, the project has specified a repair procedure. This procedure can be divided into two basic areas. These are:

A) Construction of the "foundation pit" and reinforcement of the foundation structures

B) Reinforcement of the front wall between the pilasters

Let us now say something closer to these two types of construction.

A) Construction of the "foundation pit" and reinforcement of the foundation structures

The excavation was secured by means of the so-called Berliner wall, with the construction timber put between HEB 120 beams with strutting in connection to the existing building construction. The size of the pit was made as small as possible, but with regard to the extent of the work performed. The depth of the support is individual in connection with the variable height of the existing or modified terrain. The drilling of piles with inserted HEB 120 beams is done with the HVS 246 rig with tools according to the current geology at the place of the borehole (spiral, hammer, ripping lip, roller, casing etc.). Riders, which are slender elements that transfer the load to the depth of the foundation soil, has been done in connection with the existing foundation.

Riders realization – HEB reinforcement is inserted into the borehole with a diameter of 220 mm and the borehole is filled from the bottom with a grouting needle with activated cement suspension IM30 with a pressure of up to 0.6 MPa – up to the height specified by the project. This is followed by gradual excavation and installation of wooden support between the steel beams. Together with the earthworks, strutting against the existing building according to the PD is done immediately with timbers of 140 x 140 mm.

Subsequently, the deep foundation for the newly formed structure linked with the existing structure will be done to reinforce the connecting corridor. Due to its weight, this will be supported by four new micropiles. Like the rider, the micropile is a slender element that transfers the load to the depth of the foundation base. With regard to space, the micropiles will be done by the LUMESA folding drilling rig. Due to the expanded existing foundation, a diamond pre-drilling will be done first in order to ensure as little damage to the existing structure as possible. A deep borehole with a diameter of up to 156 mm follows, immediately provided with a 76/10 mm diameter steel pipe with a smooth and sleeve part for grouting. Afterwards, the borehole is filled with activated cement suspension with a grouting needle from the bottom up to the mouth of the borehole. After the grout sets and starts to harden, grouting is carried out by means of a depth obturator at individual levels at a distance of 0.5 m by pressure up to 4.5 MPa. The micropile roots are finished in this construction and made in solid to solid powdery/sandy soil.

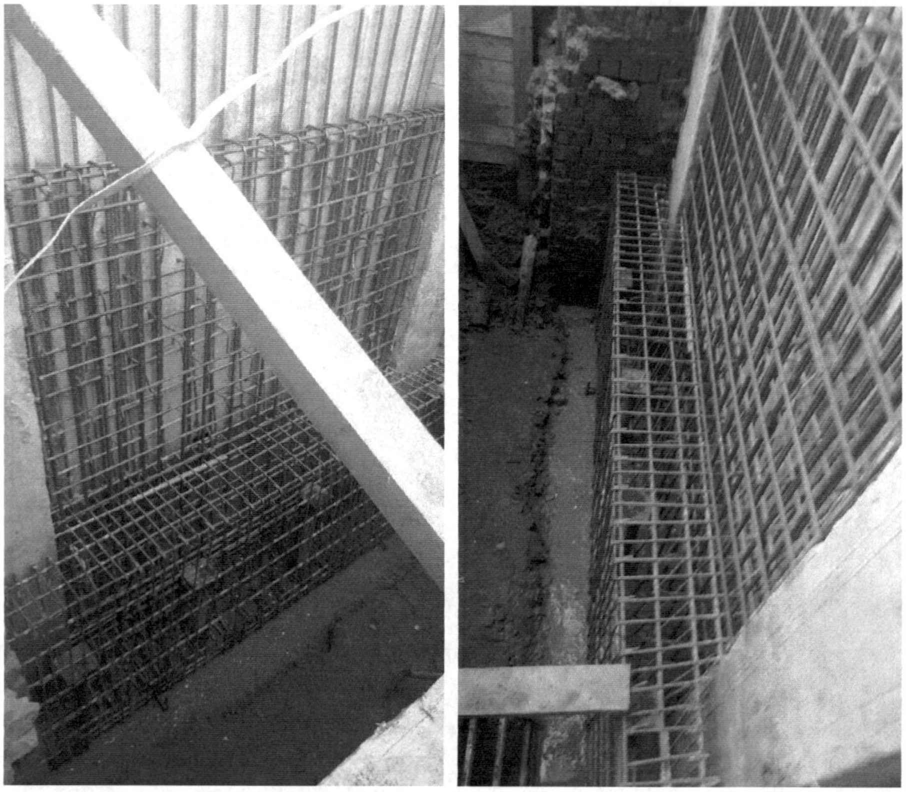

Figure 3. Two views of the reinforcement of the foundation structures before the concrete mixture was poured.

This new monolithic structure is placed on an expanded original foundation and linked to the foundations via bolts both horizontally and vertically and also supported on the micropiles. The reinforcement is cascading – the first cascade is 1500 mm high and 700 mm wide, with length always between the wall pillars. The second part of the cascade with a height of -200 mm below the modified terrain and a width of 300 mm is next; the reinforcement of the front wall continues. Standard concrete reinforcement of various diameters according to the PD, with proper interconnection of all parts of the structure from the back to the face, is used for the reinforcement. Vertical bars for reinforcing the wall above the terrain are raised from the second part of the cascade. R10 anchors in the new structure space, bonded in the connecting corridor openings with a two-component chemical adhesive, are used for linking the new and the original parts. The structure of the new part is designed from cast concrete of class C30/37 – XC2 into one-sided formwork.

B) Reinforcement of the front wall between the pilasters

After completing the previous steps, the construction pit is filled to the level of new wall reinforcement at the base. Compaction with plate compactors / rammers is a matter of course. For further work, it is necessary to erect a stable tubular scaffolding at the connecting corridor structure – this type, in combination with the system type, is probably the most variable. For further work, it is necessary to have appropriate work surfaces along the entire height of the connecting corridors. The scaffolding is used to bush hammer the surface of the structure as a basic step for the subsequent better surface blasting. The surface must therefore be thoroughly cleaned of incoherent parts of concrete, plaster, paint and completely roughened before the rehabilitation is started. These properties are achieved in the structure by a high-pressure water jet with a pressure of 2000–2500 bar. The blasting must be monitored at all times and the pressure must be regulated with regard to the quality of the existing concrete. After blasting, the exposed reinforcement is rehabilitated. It is properly cleaned and chemically treated against corrosion. In case of high corrosion, any reinforcement loss is checked, and it is subsequently passivated (protected) by prescribed materials. At this moment, the entire structure is being checked, the technical condition is generally assessed with respect to all damage – i.e. imperfections, ballast beds and missing reinforcement. There is grouting of the ballast beds with the EBL machine with pressures of up to 1.3 MPa using packers with finely ground Portland microcement.

In the case of missing reinforcement, namely horizontal or vertical bearing reinforcement or stirrups on pillars, the wrapping must be carried out due to the transfer of stress from one reinforcement to another, and no less important is a deviation in case of the vertical reinforcement, where the reinforcement is exposed to horizontal alternating effects of tension and stress load due to wind. The wrapping consists in bonding the composite reinforcement made of carbon fabric and linked from the back to the face using strips and M10 threaded rods provided with washers and nuts. In the upper part, or more precisely above the terrain up to the vent gaps at the roof of the connecting corridor, it is a linked structure reinforcing the existing wall between the pillars with a final thickness of 140 mm made by the application of shotcrete of strength class C30/37. The total thickness consists of two technological spraying steps. The first one is the concrete reinforcement connected to the reinforcement from the foundations with proper lap joint according to the PD. The reinforcement is generally supplemented with KH30 welded wire mesh made of 6 mm wire and 100 x 100 mm meshes. Laying of the mesh is prescribed by the PD and it is necessary to lap joint by at least 2 meshes when connecting.

The first spraying layer is interconnected with the existing wall via R10 reinforcement anchors bonded with a two-component chemical adhesive and bound to the KH30 mesh. The reinforcement is further reinforced by closing edge profiles lining both sides of the columns and linked to these pillars by a flat plate bonded with a two-component adhesive. The first spraying layer is done in height with a bagged mixture by a dry shotcrete technique. The technique uses the SSB 05 machine and an air high-pressure compressor with a supply of water to the spraying nozzle. The application of the mixture is carried out with a spray gun from a distance of 0.6–1.5 m from the substrate by a slight circular spiral movement so as to create a uniform layer of concrete. The water coefficient is considered in the range of 0.4–0.6, and this ratio and the entire spraying quality are influenced by the worker's experience. This is the reason for mixing the water into the dry mixture in the gun at the nozzle by the movement of the valve by the worker – it is therefore necessary to allocate the most experienced team members for this work. The amount of fallen material in bagged (very dry and fine) mixtures is up to 20%. The fallen material must not be used again.

The second layer is a surface layer. It is reinforced with composite mesh of basalt reinforcement with fibre diameter of 3 mm and meshes of 100 x 100 mm. Thanks to its technological base, the material is non-corroding, the anchoring of the mesh is done using cold-drawn helical stainless-steel reinforcement. The composite is, with a sufficient protection, linked to these helical anchors with a diameter of 4.5 mm, which are bonded in the structure with a two-component chemical adhesive. The composite reinforcement serves as a reinforcement layer for shotcrete, which is intended as shrinkage prevention, however, it does not serve for static effect of the linked structure. The final bearing layer of the linked structure follows, which is, due to the application of further technological steps of the façade, adjusted to prescribed shapes by tearing.

Figure 4. On the left already complete reinforcement of the foundation before the wall reinforcement, on the right a realization scaffold for the installation of wall reinforcement - KARI and polymer networks.

Figure 5. Anchor reinforcement and KARI nets between pilasters ready to apply shotcrete.

Figure 6. Left - made polymer mesh anchored by helical reinforcement before application of sprayed concrete, right - made wrapping and reinforcement of the structure by steel profiles and carbon slats

6. Conclusion

Concrete structures are very demanding not only for the design but also in terms of the execution phase. We must not forget about the possible emerging risks during the implementation work – it is difficult to determine which activity is less or more important. Every designer must look at the work with the utmost caution and humility, the client must not save at the expense of the work in case of complex structures, but must cooperate with the designer within the framework of the designer's supervision, who personally overlooks the work. When making any concrete structures, it is necessary to strictly observe the PD, technological discipline and check the supplied materials to minimize future defects, concrete degradation, durability and safety of the whole structure.

References

[1] J. Všetečka, Author's photo archive, 2019.
[2] M. Ďubek, M. Bederka, and M. Grinč. "Identification of the position reinforcement embedded in concrete," *Buildustry,* number 4/2017, pp. 28-31, 2017.
[3] M. Ďubek, P. Makýš, S. Ďuběk, and M. Petro. "The Evaluation of the Content of Fibers in Steel Fiber Reinforced Structures and Image Analysis," *Journal of civil engineering and management,* number 24, pp. 183-192, 2018.
[4] R. Drochytka, J. Dohnálek, J. Bydžovský, V. Pumpr, A. Dufka, and P. Dohnálek, *Technical conditions for rehabilitation of concrete structures*, 2012.
[5] ČSN 73 1201. *Design of concrete structures of buildings,* Prague, The Institute of Technical Standardization, Metrology and State Testing, 2010.
[6] ČSN EN 1990. *Eurocode: Basis of structural design*. Prague, The Institute of Technical Standardization, Metrology and State Testing, 2004.

IOP Conf. Series: Materials Science and Engineering **728** (2020) 012012 doi:10.1088/1757-899X/728/1/012012

Level of Total Volatile Organic Compounds (TVOC) in the context of Indoor Air Quality (IAQ) in Office Buildings

Michal Kraus [1], Ingrid Juhásová Šenitková[1]

[1] Institute of Technology and Business in České Budějovice, Department of Civil Engineering, Okružní 517/10, 370 01 České Budějovice, Czech Republic

info@krausmichal.cz

Abstract. This paper is focused on the quality of the indoor environment in terms of Total Volatile Organic Compounds (TVOC) in the office buildings. Working in the office is one of the most common occupations in developed countries of the world. Employees spend up to 25% of their time in the workplace each year. Obviously, the internal environment affects not only the health of workers but also their performance and productivity. Observe the long-term level of Total Volatile Organic Compounds in the context of Indoor Air Quality (IAQ) in a single cellular office is subjected within the presented study. IA-Quest (Indoor Air Quality Emission Simulation Tool) developed by the National Research Council is used for emission of VOCs from building surface materials and furnishings in time. The simulation and monitoring of TVOC concentration are carried out in two phases. In the first phase, only building materials such as walls, doors, floor, ceiling, window, etc. are considered. In the second phase, the basic furniture including the office table and chair, coffee table, 2 chairs and wardrobe are included in the calculation. TVOC concentration values in a furniture room are 10 - 15% higher than in a room without furniture.

1. Introduction

People spend up to 90 percent of their time in buildings. A quality indoor environment is therefore very important. Poor and inadequate quality is linked to a number of health problems of building users [1]. In exceptional cases, it may also be related to premature morbidity/mortality [2,3]. Overall exposure levels of pollutants inside buildings are usually higher than outdoor. One of the key objectives of any commercial office building is to provide a place for people to work. Working in the office is one of the most common occupations in developed countries of the world. The usual standard working time is 8 hours per day, i.e. 40 hours a week. From this, it can be concluded that people spend 20 - 25% of their time at work every year.

Indoor air quality (IAQ) in office buildings not only has a major impact on the performance and productivity of users but especially on their health. Indoor air quality in working places is generally considered as one of the most reliable environmental risks to human health [4]. Prolonged stay in spaces with high levels of pollutants is associated with Sick Building Syndrome (SBS). The quality of the indoor environment in office buildings is addressed by a number of studies around the world [4-8]. One of the significant symptoms of SBS is volatile organic compounds that have been emitted into the interior air for a long time. Sources of volatile organic compounds in offices include furniture, wall and flooring, photocopiers, printers cleaning products, etc. Short-term exposure to the high level of TVOC could cause headaches, dizziness, drowsiness, nausea, eye and respiratory irritation or allergic reactions.

The volatile organic compounds that are present in indoor air include carcinogenic, harmful and allergenic substances. Commonly occurring indoor VOCs are ethylbenzene, carbon tetrachloride, benzene, 1,1,1-Trichloroethane, xylenes, tetrachloroethylene, toluene, trichloroethylene, styrene, 1,4-Dichlorobenzene. Previously, the typical representative was formaldehyde. According to Lars Mølhave [9], if the TVOC concentration is less than 0.2 mg/m^3 then users are not exposed to irritation and negative effects. Most users experience a feeling of discomfort when the limit value of 3 mg/m^3 is exceeded. Headache and other neurotoxic effects occur when the TVOC concentration is higher than 25 mg/m^3.

2. Basic requirements for office buildings and office rooms

The basic requirements for offices are given in CSN 73 5305 Administrative buildings and rooms. This standard specifies principles for the design of office buildings and premises, building complexes or parts thereof that contain premises for administrative, conceptual and managerial activities. Workplaces in office buildings can be organized as cellular or open space system. Minimum spaces are set in dependence on the activities and equipment of the office. Minimum office space without meeting space and storage space is 5 m^2 (8 m^2 recommended). The minimum office area without space for dealing with the shelf is 8 m^2 (10 m^2 recommended). Minimum office space with meeting space, without shelf, is 10 m^2 (12 m^2 recommended). Minimum office space with meeting space and shelf space is 12 m^2 (16 m^2 recommended). The minimum ceiling height is 2700 mm. The recommended height is 3000 mm.

The minimum working surface of the table should be at least 750 x 600 mm. The latest trend is a height-adjustable desk table that allows both sitting and standing work. The depth of the table is 800 mm, in some situations 600 mm is accepted. The height of fixed office desks for sitting should be 740 ± 20 mm.

Most offices are currently mechanically ventilated using an air conditioner. Table 1 shows the optimum value of ventilation intensity rate according to Ashrae and EN 15251 for the office room with the floor are 10.0 m^2. The minimum supply air must be increased at additional loads such as heat or odors.

Table 1. Ventilation intensity rate according to ASHRAE and EN 15251.

		l/s per person	l/s per m^2	l/s	m^3/h per person	m^3/h per m^2
ASHRAE		2.50	0.30	5.50	19.80	1.98
EN 15251	**I**	10.00	1.00	20.00	72.00	7.20
	II	7.00	0.70	17.00	61.20	6.12
	III	4.00	0.40	8.00	28.80	2.88

Noise from office and related activities should not exceed 35 - 45 dB. For open offices, the noise limit is 50 dB. For undemanding work, the limit can be increased to 60 dB. Daylight and artificial lighting should be combined. Daylight should prevail. The office room should be illuminated evenly. The light intensity should be about 500 lx.

3. Research Methods

According to the information above, the standardized optimum size of the single cellular office with 1 workplace, put off area and cabinets is 10.0 m^2. The ground plan dimensions are 3.200 x 3.125 meters. The clear height of the room is 3.0 meters. The floor plan of the experimental single cellular office is shown in Figure 1. The air volume is 30 m^3 without taking into account the equipment and furnishing. The window is solid, non-opening, 4.5 m^2. Ventilation is forced.

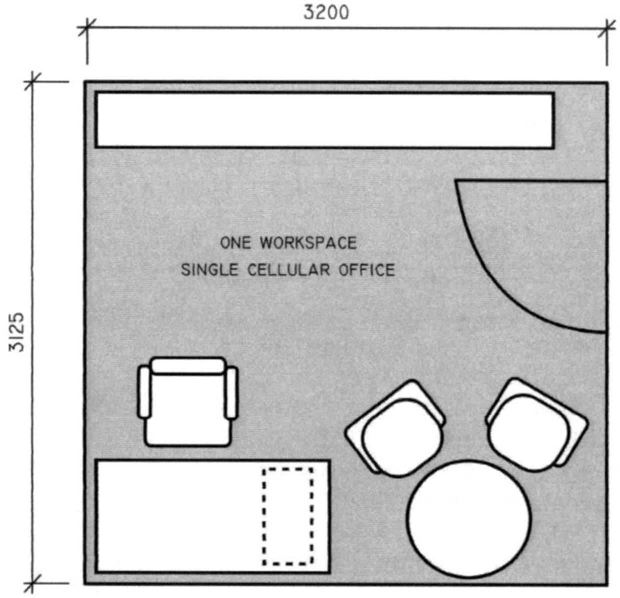

Figure 1. Floor plan of the observed single cellular office.

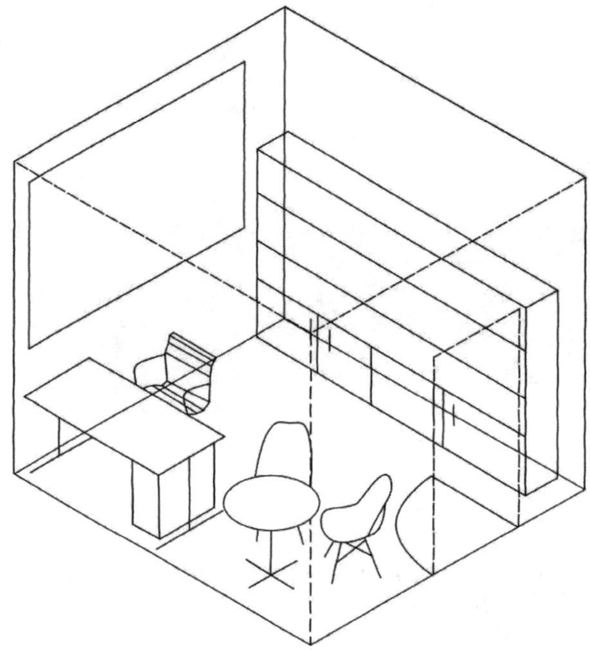

Figure 2. Simple axonometry of the observed single cellular office.

A commercial nylon carpet suitable for office use is chosen as flooring. The ceiling is made of gypsum wallboards (15.9 mm) with fire protection. The walls are treated with interior painting. The basic material characteristics of the experimental single cellular office room without equipment are given in Table 2. The table contains information about the material, emissions surface, surface area and range emission factor.

Table 2. Material characteristics of the experimental office room without furnishing.

Structural unit	Material	Emission surface [m^2]	Range emission factor [$l/s.m^2$]
Floor	Commercial nylon carpet	10.00	1.103
Ceiling	Gypsum wallboard, fire-code	10.00	3.866 – 5.232
Walls	Interior paint	31,85	55.287 – 154.450
Door	Particle board	1.60	3.866 – 5.232
Windows	Glass	4.50	-

Basic office equipment and furnishing include 1600 mm x 800 mm desk, 1x pedestal under the desk, 1x task chair, coffee table, 2x chairs, and cabinet (3000 x 2000 x 500 mm). The basic material characteristics of the experimental single cellular office room including equipment and furnishing are defined in Table 3.

Table 3. Material characteristics of the experimental office room including furnishing.

Structural unit	Material	Emission surface [m^2]	Range emission factor [$l/s.m^2$]
Floor	Commercial nylon carpet	10.00	1.103
Ceiling	Gypsum wallboard, fire-code	10.00	3.866 – 5.232
Walls	Interior paint	31,85	55.287 – 154.450
Door	Particle board	1.60	3.866 – 5.232
Windows	Glass	4.50	-
Furnishing	Particle board	22,75	3.866 – 5.232

Ventilation rate time schedule is illustrating in Table 4. Category II according to EN 15 251 is considered. According to Table 1, the ventilation rate is 61.20 m^3/h during working hours, i.e. from Monday to Friday from 06:00 to 18:00. The minimum allowable air exchange, 0.10 m^3/h, is maintained between 18:00 and 06:00 on weekdays. Minimum air exchange is also considered on weekends and public holidays.

Table 4. Ventilation rate time schedule.

	06:00 – 18:00	18:00 – 06:00
Working days (Monday – Friday)	61.20 m^3/h	0.10 m^3/h
Weekend (Saturday – Sunday)	0.10 m^3/h	0.10 m^3/h

The Indoor Air Quality Emission Simulation Tool calculates indoor air concentrations using a mass balance approach as shown in the following equation:

$$V\cdot\frac{dC}{dt} = Q\cdot C - Q\cdot C_{in} + \sum^{n} EF_i\cdot A_i \qquad (1)$$

where C is VOCs concentration [mg/m^3], V is air volume [m^3], Q is ventilation rate [m^3/h], C_{in} is VOCs concentration in ventilation air [mg/m^3], A_i is surface area of the i_{th} source [m^2], EF_i is the emission factor of the i_{th} source [m^2], and n is the number of sources [3, 10].

4. Results and discussions

The estimated TVOC emission rate using Indoor Air Quality Emission Simulation Tool (IA-QUEST) are illustrated in Figure 3 and Figure 4. Figure 3 shows the expected long-term TVOC concentration over a period of 1 year (8 760 hours) for the ventilation rate time schedule as is defined above (Table 4). In the first phase, only building materials are considered without the influence of equipment and furniture. Basic statistical analysis is shown in Table 5. The mean value is 11.16 mg/m^3 with the standard error of 0.457. The median of TVOC concentration is 2.06 mg/m^3. The minimum value is 0.09 mg/m^3 and the maximum value is 874.37 mg/m^3. The rest of the basic statistics are shown in Table 5.

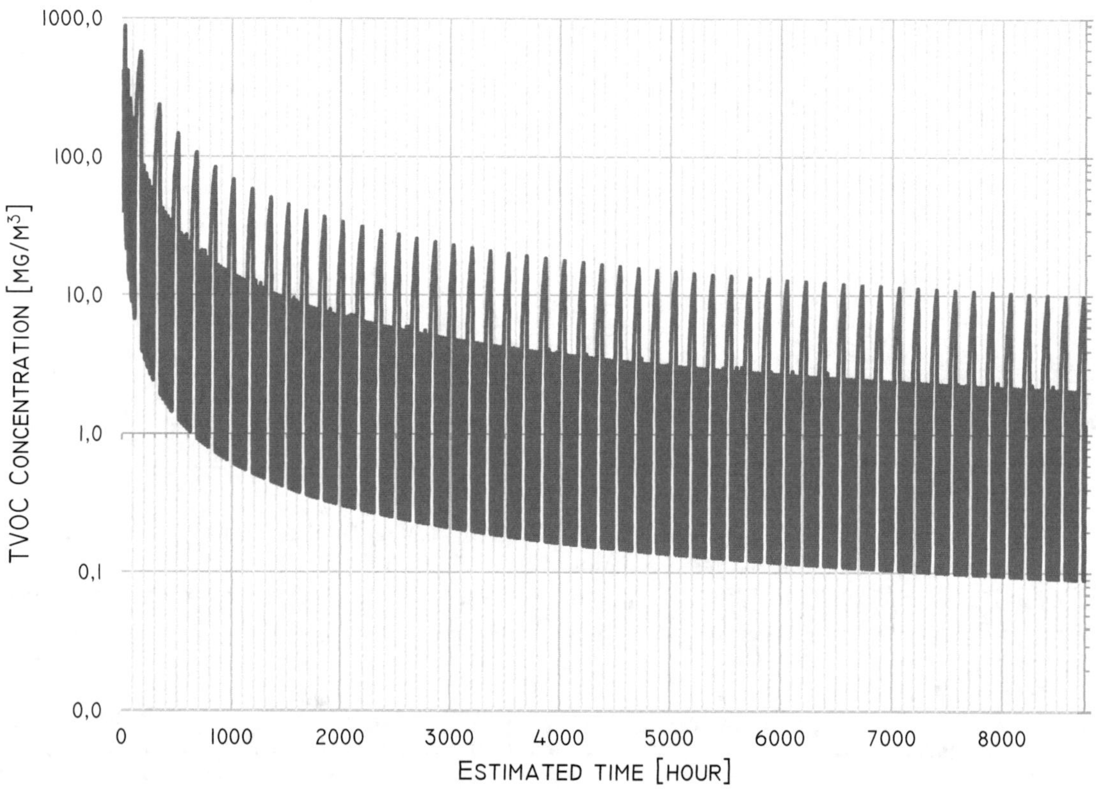

Figure 3. Long-term TVOC concentration in the single cellular office without furnishing.

Figure 4 illustrates long-term TVOC concentration in the single cellular office with the impact of the furnishing. The influence of electrical appliances is not considered. The study [11] deals with the influence of electrical appliances on indoor quality and TVOC emissions. The mean value is 12.65 mg/m^3 with standard error of 0.504. The median of TVOC concentration is 2.30 mg/m^3. The minimum value is 0.10 mg/m^3 and the maximum value is 908.58 mg/m^3.

The long-term effect of furniture is relatively low. The average addition of furniture to TVOC concentrations is about 13%. In the case of offices without furniture, TVOC concentration of less than 3.0 mg/m^3 is achieved for the first time in 233 hours (9 days). A concentration of less than 0.2 mg/m^3 is first achieved in 3203 hours (133,45 days). In the second case (with furniture), TVOC concentration of less than 3.0 mg/m^3 is achieved for the first time in 274 hours (11,4 days). A concentration of less than 0.2 mg/m^3 is first achieved in 3709 hours (154,54 days).

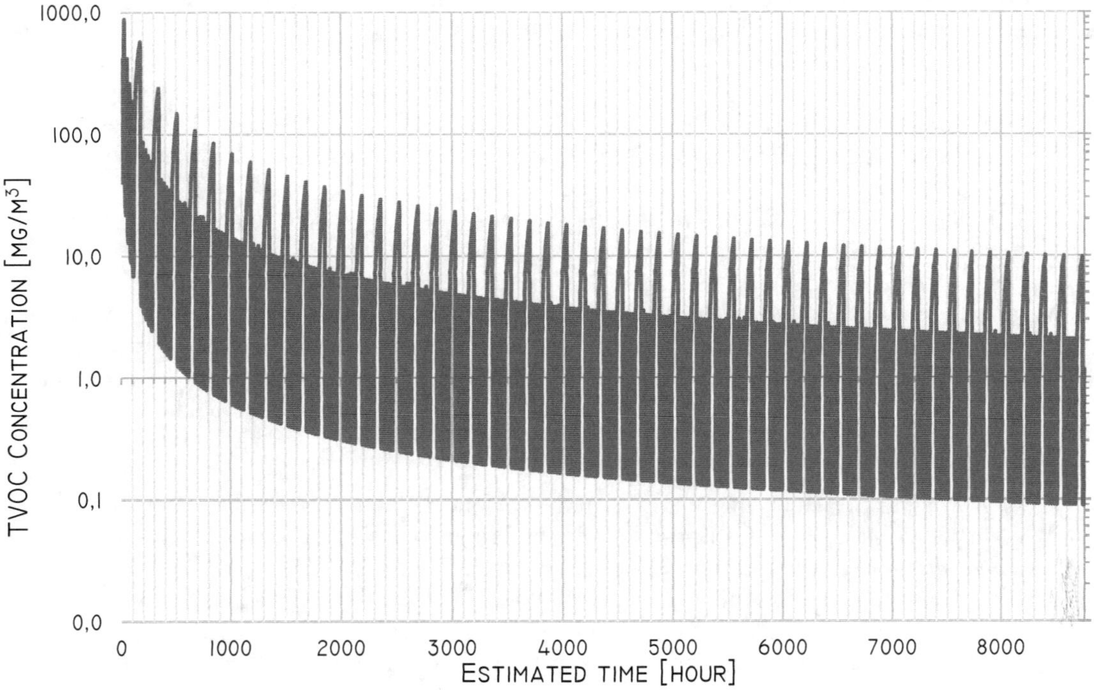

Figure 4. Long-term TVOC concentration of the single cellular office with furnishing.

Table 5. TVOC concentration without/with furnishing.

	TVOC concentration without furnishing	TVOC concentration with furnishing
Mean	11.1558	12.6516
Std. Error of Mean	0.45708	0.50377
Median	2.0600	2.3100
Mode	0.10	0.11
Std. Deviation	42.77997	47.15025
Variance	1830.126	2223.146
Skewness	9.795	9.430
Std. Error of Skewness	0.026	0.026
Kurtosis	119.667	110.139
Std. Error of Kurtosis	0.052	0.052
Range	874.28	908.48
Minimum	0.09	0.10
Maximum	874.37	908.58

5. Conclusions

Ensuring satisfactory indoor air quality (IAQ) and indoor environmental quality (IEQ) is one of the basic principles of sustainable development because people spend a major part of lifetime indoor. The level of TVOC concentration for one year in the single cellular office is demonstrated in the contribution. Based on the prediction, it is clear that the values of concentrations of volatile organic compounds exceed the permitted limits. Acceptable TVOC concentrations will be achieved approximately one-third of the year after commissioning the building. Office equipment increases concentration of VOCs about 10%. Reduction of TVOC concentration is necessary to prevent Sick Building Syndrome (SBS). The main aim of sustainable development is to ensure the optimum and suitable quality of the internal environment with minimum operational and investment costs.

References

[1] C. Shrubsole, S. Dimitroulopoulou, K. Foxall, B. Gadeberg, and A. Doutsi, "IAQ guidelines for selected volatile organic compounds (VOCs) in the UK," *Building and Environment*, vol. 165, 106382, 2019.

[2] F. J. Kelly, and J. C. Fussell, "Improving indoor air quality, health and performance within environments where people live, travel, learn and work," *Atmospheric Environment*, vol. 200, pp. 90-109, 2019.

[3] M. Kraus, and. I. Juhásová Šenitková, "Material VOC Emissions and Indoor Air Quality Simulation," *IOP Conf. Series: Materials Science and Engineering*, vol. 603, 052082, 2019.

[4] O. J. Adebayo, O. Abosede, F. B. Sunday, A. A. Ayooluwa, A. J. Adetayo, S. J Adeola, and A. F. Alaba, "Indoor Qir Quality Level of Total Volatile Organic Compounds (TVOCS) in University Offices," *International Journal of Civil Engineering and Technology*, vol. 9, issue 11, pp. 2872-2882, 2018.

[5] J. M. Daisey, A. T: Hodgson, W.J. Fisk, M. J. Mendell, and. J. T. Brinke, "Volatile organic compounds in twelve California office buildings: Classes, concentrations and sources, *Atmospheric Environment*, vol. 28, issue 22, pp. 3557-3562, 1994.

[6] M. Ongwandee, R. Moonrinta, S. Panyametheekul, Ch. Tangbanluekal, and G. Morrison, "nvestigation of volatile organic compounds in office buildings in Bangkok, Thailand: Concentrations, sources, and occupant symptoms," *Building and Environment*, vol. 46, issue 7, pp. 1512-1522, 2011.

[7] L. E. Ekberg, "Volatile organic compounds in office buildings," *Atmospheric Environment*, vol. 28, issude 22, pp. 3571-3575, 1994.

[8] C. R. de O. Nunes, B. Sánchez, C. E. N. Gatts, C. M. S: de Almeida, and M. C. Canela, "Evaluation of volatile organic compounds coupled to seasonality effects in indoor air from a commercial office in Madrid (Spain) applying chemometric techniques," *Science of The Total Environment*, vol. 650, part 1, pp. 868-877, 2019.

[9] L. Mølhave, "Volatile Organic Compounds, Indoor Air Quality and Health," *Indoor Air*, vol. 1, issue 4, pp. 357-376, 1991.

[10] D. Won, *IA-QUEST (Indoor Air Quality Emission Simulation Tool),* 2008

[11] I. Juhásová Šenitková, and M. Kraus, "Indoor TVOC and odor pollution – Chemical and sensory assessment using the glass test chamber," *JP Journal of Heat and Mass Transfer*, vol. 15, issue 3, pp. 653-673, 2018.

IOP Conf. Series: Materials Science and Engineering **728** (2020) 012013 doi:10.1088/1757-899X/728/1/012013

Use of Information Model in Creating the Investment-Construction Project

V M Cherkina[1], **N S Shushunova**[1]

[1]Moscow State University of Civil Engineering, Department of Integrated Safety in Building, 129337, 26, Yaroslavskoye Shosse, Moscow, Russia

nshushun@gmail.com

Abstract. The use of information modeling technology in creating the investment-construction object provides a number of advantages to design, construction and operating organizations. Leading development companies should be transformed into a kind of IT-structure. In this regard, construction companies that develop their own strategies for switching to digital technologies and generate digital assets should, first of all, create the basis for their functioning - the integrated information model of the object - the BIM model. The latter includes all sections of the project and contains reliable and relevant information about the existing physical object. The concrete examples in this article show the results of the successful implementation of this approach in relation to capital construction projects of varying complexity.

1. Introduction

With the development of Building Information Modeling or BIM technologies, it becomes clear that the focus of interest in the construction industry is shifting toward the consumer, communication with him, as well as the means, channels and tools that make this communication as efficient and comfortable as possible. Developers in America and many European countries are already actively investing in startups which are creating digital ecosystems in residential projects and contribute to the digital transformation of the entire industry. Information modeling processes allow you to integrate and link the individual elements of technology, organize the collection and processing of data, support the process of making managerial decisions.

The relevance of using information modeling is also confirmed by the fact that since July 2019, Russian developers working under the shared construction scheme are required to apply to bank project financing exclusively using escrow accounts, i.e. with the deposit of monetary amounts from a third party so that these amounts are issued to the developer after fulfillment of certain obligations by him. This complication of financing instruments brings to life such qualities of the project as its transparency, accessibility to all participants in the transaction on-line, the ability to quickly analyse the sensitivity of the model to variations in design parameters, etc. In addition, there is a need for additional control of environmental friendliness, safety, quality of construction and ensuring the predictability of successful completion of the project. All these issues are extremely difficult to solve in the complex without using the BIM-model of the object.

2. Methodology

There are standards and National Building Specification (NBS), UK-based technology platform and system of construction specification, used by architects and other building professionals, that help organize information modeling processes and establish modes for using information models at various stages with multiple methods for geometric and energy-related data acquisition to achieve one or more project goals [1,2].

2.1. Determination of the required level of study at various stages of the project

The factors, determining the necessity of buildings refurbishment are related to the problems encountered in particular building. We are considered one of the tasks of the design process - determining the required level of development of model elements at various stages of the project, in other words, determining the maturity level of information modeling technology.

Figure 1 shows the maturity model of information modeling technology, showing progress from 2D CAD to BIM Level 3. The model describes the maturity levels in relation to the ability of construction industry organizations to produce, store and share structured information.

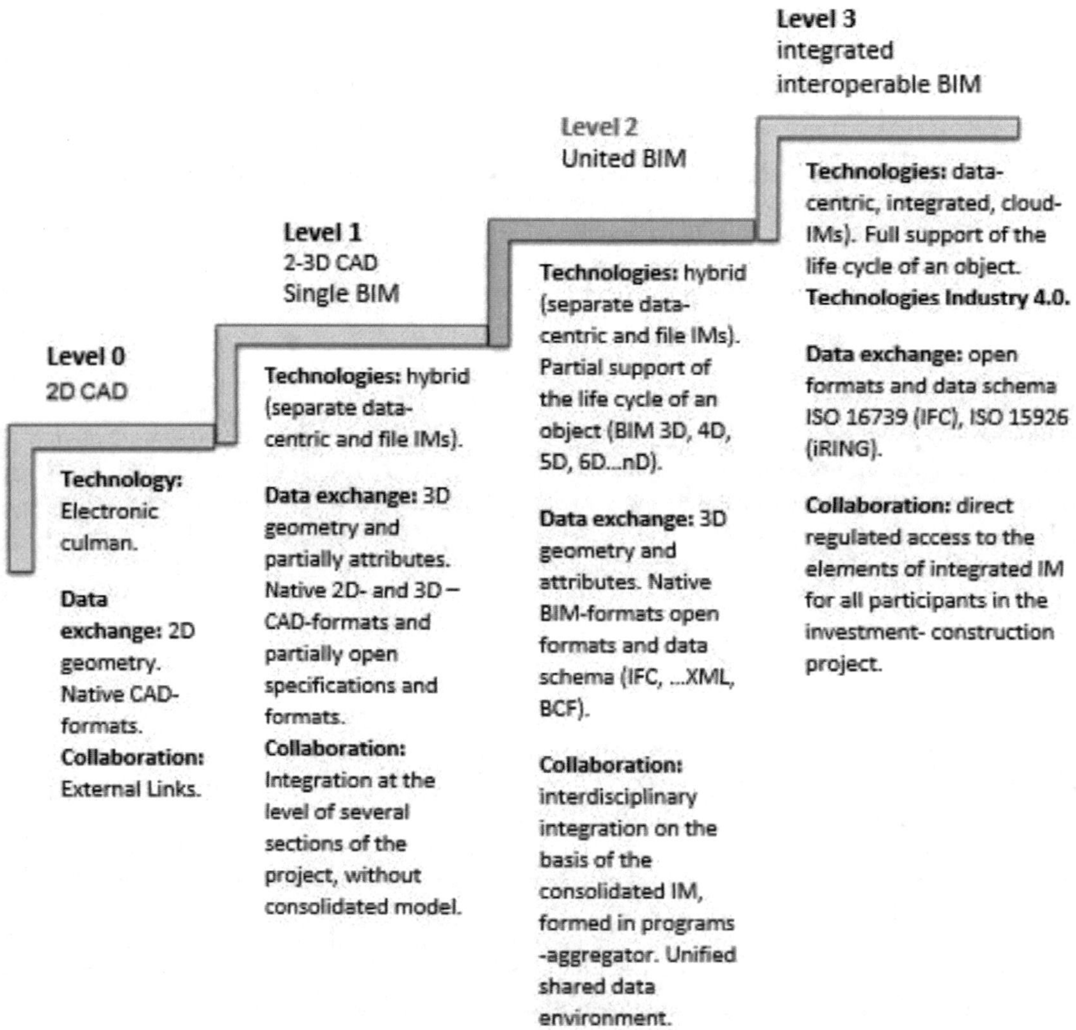

Figure 1. Maturity model of information modeling technology.

Level 0: Uses traditional CAD in 2D format. Data exchange is carried out mainly at the level of 2D geometry. Collaboration is practically absent or implemented through external links.

Level 1: This is the level at which most organizations currently operate. Usually this is a combination of 3D CAD / BIM and 2D CAD (preparation of design and working documentation). A composite model is not formed, as many disciplines work in 2D. Data is exchanged at the level of 3D geometry and attributes within disciplines using 3D CAD / BIM solutions.

Level 2: This is the current maturity level of BIM technology. The main difference from the previous levels and the main goal of this level is the organization of joint coordinated work of multidisciplinary project teams based on a consolidated model placed in a shared data environment. Data exchange is carried out at the level of 3D geometry and attribute information. This level involves the addition of the following measurements: 4D (time, linking the model to calendar and network charts) and 5D (cost, purpose and linking of resources and prices), as well as partial use of the BIM model at all stages of the object's life cycle.

Level 3: It involves working through web services of all participants in an investment and construction project, including all design disciplines, with a single integrated data-centric information model based on open schemes, data formats, semantics and ontologies. Note that currently there are no implementations of documents and international standards at this level.

In the framework of 3D modeling, we can talk not only about building models of objects, but also about filling them with data, which in turn allows us to optimize management decision-making processes and subsequently connect product design tools with their production tools.

The determination of the maturity levels of information modeling technology can be used in the formation of technical specifications for the creation of a BIM model of an object [3,4].

Some scientists bring together the theory of Life Cycle Assessment (LCA) and the capabilities of BIM to survey the current developments in the energy efficiency of structural systems [5].

3. Results and discussions

Nowadays developments in the smart energy sector focuses on the techniques in the main application areas, while highlighting some of the key challenges currently faced and outlining future pathways for the sector [6].

Consider the process of forming the technical specifications for BIM "Reconstruction of covering with landscaping system in energy-efficiency building". Covering with landscaping system belong to the field of construction, namely to the constructive solution of the coating, and can be used in the construction of “living” coatings for residential buildings, sports facilities, as well as coatings for any stations. The design provides also vertical system, as well as the ability to place and use devices that accumulate and convert energy, elements of water-irrigation control. The 3D model shows a continuous-form wall panel of such a covering of a public zone in a building (Figure 2).

Figure 2. The process of forming the technical specifications for the BIM reconstruction of the project.

The technical result is achieved by the fact that space-planning schemes for the mutual arrangement of the coverage space. Thus, the problem of non-waste installation of structures is solved, since there is no labour-intensive work associated with cutting products.

3D models are needed to define the set of vital technical parameters depending of a vast spectrum of environmental and functional influences of the existing buildings under analysis [7].

Improved safety through increased construction awareness from easy review of complex details or processes on site, out of the working digital modeling benefits. Most of the public indoor facilities conditions do not meet the requirements of the hygiene standard, the thermal comfort parameters, in this case they must be renovated.

Creating of 3D models are allows to define the variety of technical parameters. The researching of coverings by P. Bevilacqua, D. Mazzeo, N. Arcurideals, M. Carbone et al. [21-26] with an experimental investigation of the dynamic thermal characteristics of a roof situated on a university building roof in south Italy, based on EN ISO 13790 Energy performance of buildings — Calculation of energy use for space heating and cooling (Figure 3).

IOP Conf. Series: Materials Science and Engineering **728** (2020) 012013 doi:10.1088/1757-899X/728/1/012013

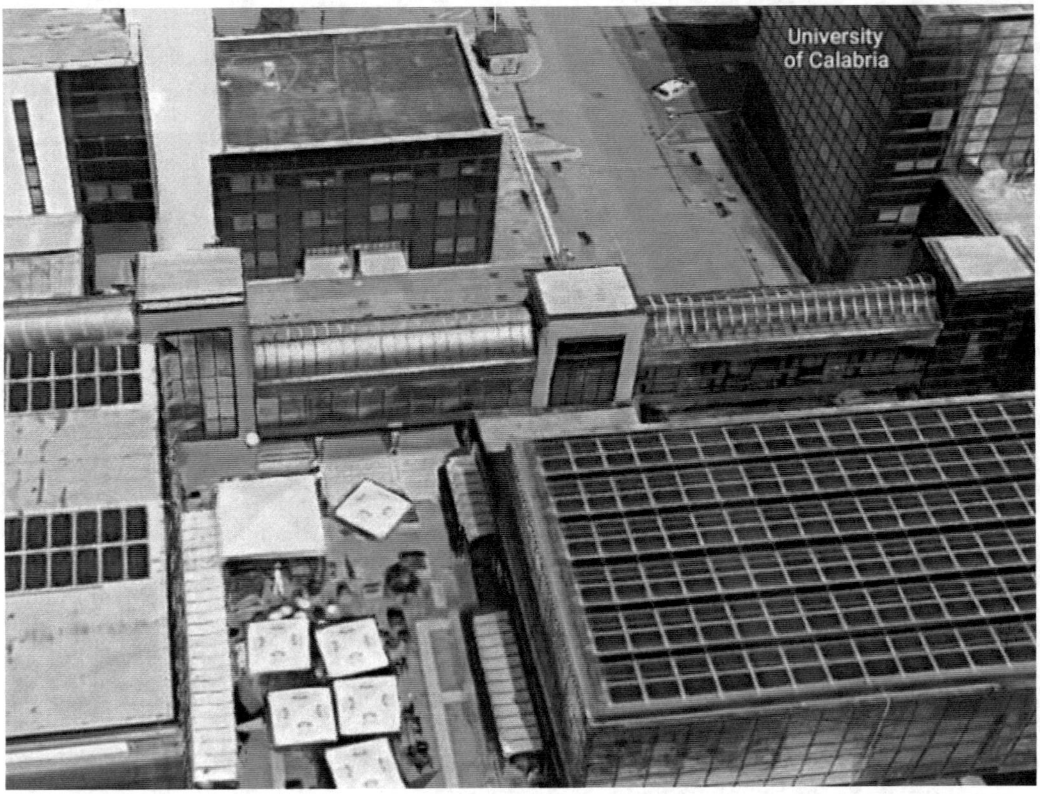

Figure 3. The 3D view of roof, situated on the University of Calabria.

There are a number of BIM implementations from around the world that can be used to guide development of Russian BIM use, especially from GSA in the USA, and COBIM in Finland, Australian National Construction Code (NCC), Building Code of Australia (BCA) [8]. Thus, IFC, or "Industry Foundation Classes", is an international standard (ISO 16739-1:2018), usable across a wide range of hardware devices, software platforms, and interfaces for many different use cases, including the risk management assessments [9, 27-29].

4. Conclusions

Nowadays many developers offer the buyer not just a specific unit of housing, but a diverse environment filled with infrastructure facilities, technological services and social activities. The choice of scenario will depend on the scale and resources of the company. However, in both cases, the priorities for investment will be BIM-design, mobile applications, intelligent systems of automatic control and management. Every third Russian developer offers high-tech solutions, which are commonly called “smart home” and “smart architecture”. BIM technologies were involved in the design of 2 million square meters. m of real estate put into operation, which is about 30% of the total fund of objects [10]. Digital transformation is beneficial for both parties. The client receives a much more convenient product, adapted to his needs, and the developer optimizes costs, increasing efficiency. A new stage in the evolutionary cycle in the development of the industry has already begun. Its main content will be the opening by developers of additional opportunities and growth points in real estate management. Both construction companies and the consumer of their products have an interest in increasing the life cycle of projects. The client will receive the most comfortable living environment saturated with services, and the developer will expand the horizon for the development of the business or its individual segments.

IOP Conf. Series: Materials Science and Engineering **728** (2020) 012013 doi:10.1088/1757-899X/728/1/012013

References

[1] *Information Modeling Guide for Customers 2019* (Case Study of Industrial Facilities), Version 1.0, Autodesk, Inc.
[2] L. Sanhudo, M. Ramos, J. Martins, R. Almeida, L. Simoes, and V. Cardoso, "Building information modeling for energy retrofitting – A review," *Renewable and Sustainable Energy Reviews*, vol. 89, pp. 249-260, 2018.
[3] J. Cepurnaite, L. Ustinovicius, and M. Vaisnoras. "Modernization with BIM technology through scanning building information," *Procedia Engineering*, vol. 208, pp. 8-13, 2017.
[4] A. A. Volkov, and L. V. Sukneva, "BIM-Technology in Tasks of the Designing Complex Systems of Alternative Energy Supply," *Procedia Engineering*, vol. 91, pp. 377-380, 2014.
[5] S. Eleftheriadis, D. Mumovic, and P. Greening, "Life cycle energy efficiency in building structures: A review of current developments and future outlooks based on BIM capabilities," *Renewable and Sustainable Energy Reviews*, vol. 67, pp. 811-825, 2017.
[6] E. O'Dwyer, I. Pan, S. Acha, "Smart energy systems for sustainable smart cities: Current developments, trends and future directions," *Applied Energy*, vol. 237, pp. 581-597, 2019.
[7] M. Li, and J. Yang, "Critical factors for waste management in office building retrofit projects in Australia," *Resources, Conservation and Recycling*, vol. 93, pp. 85-98, 2014.
[8] S. Beazley, E. Heffernan, and T. McCarthy, "Enhancing energy efficiency in residential buildings through the use of BIM: The case for embedding parameters during design," *Energy Procedia*, vol. 121, pp. 57-64, 2017
[9] *BuildingSMART 2019 "IFC Overview summary"* [Online]. Available: https://technical.buildingsmart.org/standards/ifc/. [Accessed: 25-08-2019]
[10] M. Berlovich, *RBC Real Estate, section "Opinions*", 2019.
[11] V. Borkovskaya, "Environmental and Economic Model Life Cycle of Buildings Based on the Concept of "Green Building"," *Applied Mechanics and Materials*, vol. 467, pp. 287-290, 2013.
[12] E. Palomo, "Roof components models simplification via statistical linearisation and model reduction techniques," *Energy and Buildings*, vol. 29, issue 3, pp. 259-281, 1999.
[13] O. Korol, *Constr. Mater*., vol, 6, pp. 3-15, 2015.
[14] L. Pérez-lombard, J. Ortiz, and C. Pout, "A review on buildings energy consumption information," *Energy and Buildings*, vol. 40, issue 3, pp. 394-398.
[15] I. Dunichkin, E. Naboni, A. Korobeinikova, and O. Poddaeva, "Issues of wind mode visualization and pedestrian comfort assessment when designing residential buildings on sloping territories in the Arctic zone," *SHS Web of Conferences*, vol. 64, 03013, 2019.
[16] R. Santos, A. Costa, J. Silvestre, L. Pyl, "Integration of LCA and LCC analysis within a BIM-based environment," *Automation in Construction*, vol. 103, pp. 127-149, 2019.
[17] B. Soust-Verdaguer, C. Llatas, and A. García-Martínez, "Critical review of bim-based LCA method to buildings, *Energy and Buildings*, vol. 136, pp. 110-120, 2017.
[18] H. Mousazadeh, A. Keyhani, A. Javadi, H. Mobli, K. Abrinia, and A. Sharifi, "A review of principle and sun-tracking methods for maximizing solar systems output," *Renewable and Sustainable Energy Reviews*, vol. 13, issue 8, pp. 1800-1818, 2009,
[19] R. Bruno, P. Bevilacqua, L. Longo, N. Arcuri, "Small Size Single-axis PV Trackers: Control Strategies and System Layout for Energy Optimization," *Energy Procedia*, vol. 82, pp. 737-743, 2015.
[20] M. Röck, A. Hollberg, G. Habert G, and A. Passer, "LCA and BIM: Visualization of environmental potentials in building construction at early design stages," *Building and Environment*, vol. 140, pp. 153-161, 2018.
[21] P. Bevilacqua, D. Mazzeo, R. Bruno and N. Arcuri, "Experimental investigation of the thermal performances of an extensive green roof in the Mediterranean area," *Energy and Buildings*, vol. 122, pp. 63-76, 2016.
[22] P. Bevilacqua, D. Mazzeo, and N. Arcuri, "Thermal inertia assessment of an experimental

extensive green roof in summer conditions," *Building and Environment*, vol. 131,pp. 264-276, 2018.

[23] P. Bevilacqua, D. Mazzeo, R. Bruno and N. Arcuri, "Surface temperature analysis of an extensive green roof for the mitigation of urban heat island in southern mediterranean climate," *Energy and Buildings*, vol. 150, pp. 318-327, 2017.

[24] V. Borkovskaya, "The Concept of Innovation for Sustainable Development in the Construction Business and Education," *Applied Mechanics and Materials*, vol. 475-476, pp. 1703-1706, 2013.

[25] M. Carbone, G. Garofalo, G. Nigro, and O. Piro, "Green Roofs in the Mediterranean Area: Interaction between Native Plant Species and Sub-Surface Runoff," *Applied Mechanics and Materials*, vol. 737, pp. 749-753, 2015.

[26] R. Bruni, N. Arcuri, and C. Carpino, "The Passive House in Mediterranean Area: Parametric Analysis and Dynamic Simulation of the Thermal Behaviour of an Innovative Prototype," *Energy Procedia*, vol. 82, pp. 533-539, 2015.

[27] V. Borkovskaya, E. Degaev, and I. Burkova, "Environmental economic model of risk management and costs in the framework of the quality management system," *MATEC Web of Conferences*, vol. 193, 05027, 2018.

[28] V. Borkovskaya, "Complex Models of Active Control Systems at the Modern Developing Enterprises," *Advanced Materials Research*, vol. 945-949, pp. 3012-3015, 2014.

[29] V. Borkovskaya, "Post Bifurcations of the Concept of the Sustainable Development in Construction Business and Education," *Advanced Materials Research*, vol. 860-863, pp. 3009-3012, 2013.

IOP Conf. Series: Materials Science and Engineering **728** (2020) 012014 doi:10.1088/1757-899X/728/1/012014

Application of Modeling Processes and 3D Print on Casting Molds for Concrete Furniture

Jiří Šál [1], Martin Dědič [1]

[1] Institute of Technology and Business in České Budějovice, Department of Civil Engineering, Okružní 517/10, 370 01 České Budějovice, Czech Republic

sal@mail.vstecb.cz

Abstract. The paper describes the practical use of modeling processes and 3D printing for the production of furniture, decorative and other objects. Using 3D modeling and printing on a 3D printer, it is possible to create a mold and subsequent product with a much more complex shape than would be possible with hand-made formwork or mold, or at least much simpler. Demonstration of applicability of these molds was carried out by an experimental method when the molds were poured out by cement or concrete from the fine aggregate. The forms themselves were formed from PLA filament. Experiments have shown the usability of such molds for complex concrete elements. These forms can be reused. Damage to the molds occurred during handling or cleaning, i.e. rough handling. During the filling itself and the process of cement hydration, they were not damaged. For a longer mold life, it is necessary to print in a larger percentage of filling, which, however, makes the mold more expensive. The durability of molds and better formwork is obtained by using formwork oil.

1. Introduction

The aim of the article is to verify the practical use of modeling processes and 3D printing for the production of furniture, furniture, decorative and other elements. The paper will verify whether it is possible, using 3D modeling and subsequent printing on a 3D printer, to create molds for casting from concrete or cement. Such castings could have much more complex shapes than conventional formwork. In addition, the use of PC modeling and subsequent printing can be simpler and less laborious than molding by hand or other methods.

Nowadays some companies are engaged in the production of concrete decorative elements, advertising products and the like [1].

2. Preparation of forms for printing

For modeling of forms is based on knowledge of general modeling of objects, which is used in building construction. In this case, the Inventor modeling software from Autodesk [2] was used, but any software that allows you to create three-dimensional models with exact dimensions can be used and the object can be exported in a .stl file.

When modeling a mold for casting concrete objects it is necessary to have knowledge of 3D printing and casting of concrete into classical molds. The modeling is preceded by the layout - reconnaissance [3] - of the mold on the parts so that their size corresponds to the printable three-dimensional area of the 3D printer used, and also the mold fulfills its purpose so that it can be opened after casting to remove the desired object and reuse. Figure 1 shows the form in the software and the final printed form.

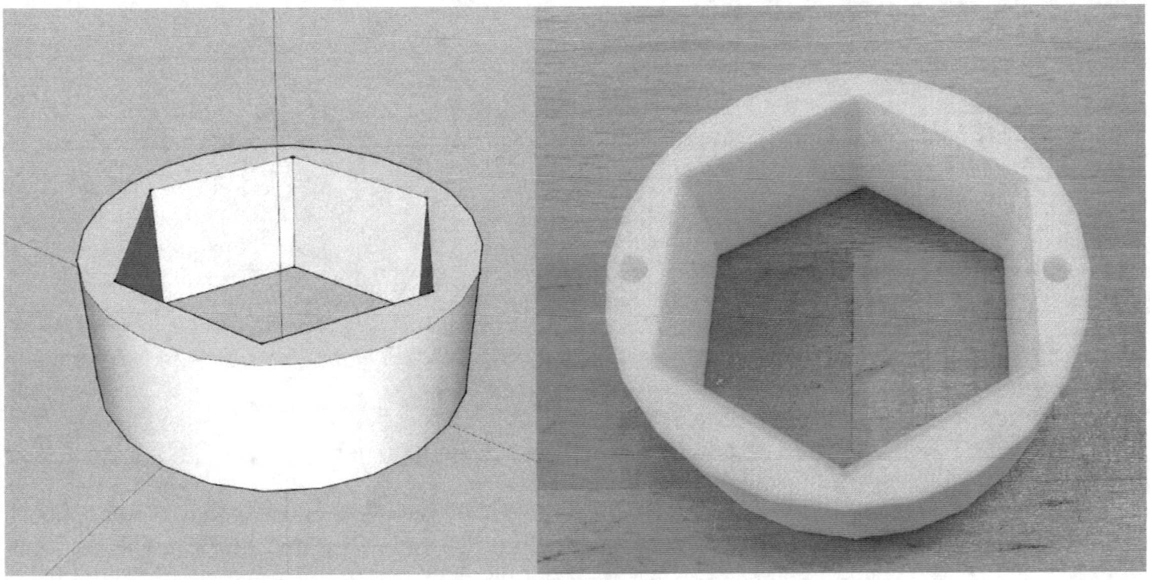

Figure 1. Form modeling in 3D software, final embossed form.

3. 3D printing of forms

Fused Deposition Modeling was used for 3D mold printing. The process of this 3D printing technology is that the thermoplastic filament is extruded through a small nozzle into fine layers that accumulate on each other to form the object. The printer heats the plastic to a molten state and pushes it (or extrudes) through a small nozzle - much like a hot glue gun, molten adhesive flows out when the trigger is pressed. The computer controls the movement of the 3D printer (based on your 3D model) and tells it to selectively deposit the molten plastic in the shape of the first layer of your object, which solidifies almost immediately as it cools. Then the next layer is extruded onto the first layer and so on. As layers build on each other - a small layer after a small layer - the physical object begins to form. [4]

The Prusa MK2 3D printer was used with the following settings:

Material:	PLA
Step:	0,2 mm
Nozzle:	0,4 mm
Perimeter:	2
Temperature:	210 °C
Filling:	20 %

For larger parts, a 0.3 mm step was used to reduce 3D printing time. A 20 % filler is sufficient for concrete molds as the mix pressure is not high enough to require a higher filler percentage. For the same reason, filament from PLA could be used.

4. Production of concrete castings

The printed form was either fixed to the base plate as shown in Figure 2 or other cases it was directly assembled from multiple pieces printed on a 3D printer. The most important thing has always been that the finished concrete product can be easily removed. At the same time, the form was undamaged and could be used further.

Figure 2. Casting of concrete into molds.

The concrete mix consisted of Portland cement mix CEM II/B-M (S-V-LL) of strength class 32.5 R according to ČSN EN 197-1 [5]. It also consists of aggregates with a 0-2 mm fraction. The aggregate grain size curve is shown in Figure 3. The mixing water was supplied with a water coefficient of 0.4 to 0.5, depending on the complexity of the mold. The more complex the form, the higher the water coefficient was used or the plasticizer was used.

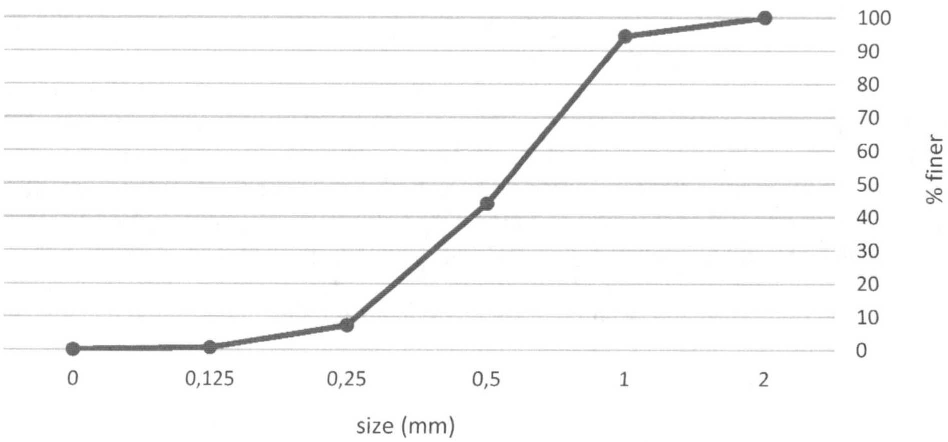

Figure 3. Grain Size Analysis of Aggregates.

The first hexagonal mold (Figure 2) was used to determine the correct ratio of binder (cement) and aggregate (silica sand). Visually the best result was 1:1, cement to sand (weight ratio). The test products are shown in Figure 4. In this figure, the final mixture is indicated by 1. The number 0 denotes the cement paste, ie without the use of aggregates.

The mold was filled by pouring the concrete mixture into the mold, printed on a 3D printer. The molds were wiped off with formwork oil. Furthermore, the sample was shaken by an impact on the table or by shaking on a vibrating table. The blend mold was covered with a plastic bag for 24 hours to prevent the blend from drying out. Then, the sample was removed from the mold and immersed in water for 24 hours to avoid cracks. According to the usual procedure, the sample should be aged under water for 28 days, but in this case, it was not necessary to measure the strength after 28 days.

Figure 4. Final castings.

5. Results and discussions, Conclusions

Experiments have shown the usability of such molds for complex concrete elements. These forms can be reused. Damage to the molds occurred during handling or cleaning, ie rough handling. During the filling itself and the process of cement hydration, they were not damaged.

For a longer mold life, it is necessary to print in a larger percentage of filling, which, however, increases the cost of mold production. The durability of molds and better form removal is obtained by using formwork oil.

A problem in the production of concrete products and decoration by this method is the limitation of the size of the printed form by a 3D printer. Each printer has certain limits in terms of print area and print height. There are two methods to solve these problems. The first method is to assemble a mold from multiple parts printed on a 3D printer and then join them together, for example gluing, screwing and the like. This method is also suitable for easier removal of the product. Thorough gluing of the individual parts does not result in visible joints on the concrete product, but it may be more difficult to remove the formwork. If parts of the mold are screwed together, they can be stripped off well, but again they are visible on the concrete product (as shown in Figure 5). Another method of eliminating the problem of mold size limitation is by mold shifting. However, this method is only suitable for some products where the mold can be moved in one direction. The manufacturing process according to this method means: 1. casting the first part into the mold; If necessary, continue from step 2. For better connection of the individual parts, the product must be still moist or need to be supplemented with wires connecting the two parts.

An example of the application of mold casting from a 3D printer is shown in Figure 5. There was created a set of decorations with ashtray, candlesticks and coaster.

Because of the formwork, it is preferable to use silicone molds. Silicone molds, however, do not retain large concrete weight for larger products. Creating a silicone mold is more difficult or less available than using 3D printing. A suitable option would be to use a combination of silicone and 3D printing.

Figure 5. Example of application of mold casting from 3D printer.

References

[1] Švec beton, *Concrete kitchens, stairs and custom furniture*, Accesible from: https://svecbeton.cz/

[2] Inventor, *Professional-grade 3D CAD software for product design and engineering*. Accesible from: https://www.autodesk.com/products/inventor/overview

[3] K. Prušková, M. Dědič, and J. Kaiser, "Possibilities of Using Modern Technologies and Creation of the Current Project Documentation Leading to the Optimal Management of the Building for Sustainable Development," *IOP Conference Series: Earth and Environmental Science*, vol. 290, issue 1, 012058, 2019.

[4] E. Lampert-Greaux, "3D Printing Basics For Entertainment Design," *Live Design*, 2018.

[5] ČSN EN 197-1 ed. 2 (722101), *Cement – Part 1: Composition, specifications and conformity criteria for common cements*, 2012.

IOP Conf. Series: Materials Science and Engineering **728** (2020) 012015 doi:10.1088/1757-899X/728/1/012015

Methodological Features of the Management of Modern Labour Protection in the Construction

E B Sugak[1]

[1]Moscow State University of Civil Engineering, 129337, 26, Yaroslavskoye Shosse, Moscow, Russia

SugakEB@mgsu.ru

Abstract. In the recent ten years, there has been a steady decline in industrial injuries and occupational diseases in the global economy. At the same time, against the background of favorable statistics on accidents, an increase in the overall incidence is observed, which are called «work related diseases». By decision of the Government of the Russian Federation to improve the production environment should ensure the reform of the management system of labour protection, focused on the prevention of the manifestation of dangerous and harmful factors. In this article there is analyse of the methodological aspects of occupational risk management, which imply a continuous and consistent cycle of activity from four stages, characterized as «the Deming-Shewhart cycle».

1. Introduction

This research is aimed at studying the methodological problems of the management of modern labour protection in the construction. Since the appearance of the first normative legal acts on labour protection, which were adopted in Germany more than 150 years ago, the labour safety system has been constantly improved, its goals and objectives have been clarified, and new methods and organizational forms have been used. Thanks to the joint efforts of government agencies, employers, labour collectives, public organizations represented by trade unions and the activities of the International Labour Organization (ILO), working conditions have improved significantly, comfort has increased, and the overall production environment has improved. As a result of the implemented measures, a steady decline in accidents and occupational diseases occurs, the number of disabled and fatal outcomes decreases.

However, simultaneously with a decrease in occupational morbidity and occupational injuries at industrial enterprises, the overall incidence began to increase progressively. Currently, non-specific for the reasons and conditions of occurrence of the disease, in the appearance and development of which the contributing mechanism or risk factor are the conditions of labour activity of people, take the leading positions in the structure of health disorders of workers. Such diseases are called "Work related diseases" (WRD). In the most cases, these are common diseases, to which the World Health Organization (WHO) refers to the following groups of diseases:

- diseases of the cardiovascular system: arterial hypertension, coronary heart disease, etc.;
- psychogenic diseases and mental disorders;
- respiratory diseases (bronchial asthma, bronchitis, cancer, etc.);
- diseases of the musculoskeletal system (pain in the spine, joints);

- violation of reproductive function.

This problem has acquired particular relevance in modern conditions. Occupational diseases do not belong to occupational diseases, while their number tends to increase with increasing length of service. It has been established that cases of loss of ability to work from WRD exceed the corresponding number of diseases in those occupational groups that do not come into contact with harmful and dangerous factors. The variety of effects on the human body of toxic substances and allergens, various physical factors, the intensification of labour leads to the transformation of existing occupational diseases and expands the range of production-related pathology. Difficulties in solving this problem are associated with the need to clearly differentiate between occupational diseases, which are an insured event and require compensation for damage to health. Production-related diseases are not, and, therefore, economic compensation for damage to the victim is not made.

The main characteristic of the multifactor production impact is the presence in the process of work of hazards and hazards of various nature and intensity, as a rule - at the level of permissible working conditions (class 2) or harmful low intensity (class 3.1). In accordance with Russian Federal Law no. 426 "On special assessment of working conditions", class 2 of working conditions means such a level of harmful and dangerous factors that do not exceed the hygienic standards and which allows the body to restore its functional state by the beginning of the next working day (shifts). Class 3.1 (hazardous working conditions of the first degree) includes such a workplace for which the restoration of an altered functional state of an employee's body occurs, as a rule, over a longer period than before the next working day (shift).

According to ILO estimates, the annual number of work-related diseases without a fatal outcome is 160 million people. And the annual mortality rate in 15 countries of the European Union is 120 thousand deaths due to the WRD, compared with 6 thousand from fatal industrial accidents. According to the ILO, there are about 100 cases of accidents due to fatal injuries in the world. Thus, the hidden risk in the loss of ability to work, in the interpretation of the ILO and WHO, is caused precisely by production-related diseases and is dominant.

2. Methodology

2.1. Total economic losses from industrial diseases and injuries

The Government of the Russian Federation in October 2011 adopted a resolution on measures to improve working conditions, preserve the lives and health of workers. The document noted that the country recorded a high level of industrial injuries, professional and production-related diseases, several times higher than in developed countries, because of unfavorable working conditions, the working time fund is significantly lost.

The government estimated the country's total economic losses from industrial diseases and injuries at 4.3% of GDP or 1.94 trillion Russian rubles, which exceeded the defense expenditures of the Russian Federation in those years. At the same time, the loss of the working time fund due to unfavorable working conditions was estimated at 1.48 trillion Russian rubles (76% of total losses), payments of the Pension Fund of the Russian Federation for early retirement pensions for work in hazardous conditions of labour were determined at 300 billion Russian rubles (15% of total losses), payments by the Social Insurance Fund for insurance provisions in connection with accidents and occupational diseases at work amounted to 159 billion Russian rubles (8% of total losses) [1]. Thus, the greatest losses of the enterprise are from labour losses due to diseases of a general nature, i.e. from production-related diseases caused by adverse working conditions.

The main problem of the negative situation with working conditions, in the opinion of the Government of the Russian Federation, is the ineffective labour protection management system, which existed in Soviet times, but which largely continues to function. It was recognized that in modern market

IOP Conf. Series: Materials Science and Engineering **728** (2020) 012015 doi:10.1088/1757-899X/728/1/012015

economy conditions it does not reduce injuries and morbidity, does not stimulate employers to improve working conditions, is not focused on preventing the manifestation of dangerous and harmful production factors, but on eliminating the consequences of occupational injuries and occupational diseases, on paying compensation to staff for work in hazardous conditions. In this regard, the Government of the Russian Federation has begun to implement measures for the transition from a compensatory model of occupational safety to a modern model based on the management of occupational risks. The new model of labour protection should allow to move away from the practice of responding to the already occurring facts of injuries or various diseases to the mechanism for developing and implementing preventive measures to prevent accidents, to preserve the health of personnel, which can ensure labour longevity [1].

2.2. The management system of organization as a complex mechanism

Any activity in the sphere of material production requires the expenditure of certain efforts aimed at ensuring its working state at the present time and in the longer term. The management system of organization is a complex mechanism using different methods and means of influencing the technical object under control and the working personnel. Management of professional risks involves a continuous cycle of activities in the sequential implementation of a set of procedures to identify and recognize negative factors of the working environment, develop measures to reduce or eliminate them, implement these measures, monitor the results and, if necessary, on corrective action. Such activity is an important part of the labour protection management system, which, in turn, should be integrated into the overall management system of any organization [2].

The famous American scientist W.E. Deming (1900 - 1993), conducting research in the field of quality management of products and services, came to the conclusion that the effectiveness of an organization's success by 94% depends on the effectiveness of the production management system and only 6% of direct actions by the artists themselves [3]. Optimization of the management system is achieved by continuous improvement of the activities of top and middle management, and the production itself is perceived as a whole system, and not as separate elements of it, which may seem incomprehensible or paradoxical in a single state.

The activity of any organization is associated with risk, therefore risk management, or risk management, must be used in all business processes. Thus, the procedures for assessing occupational risks are recommended as an initial stage in the formation of approaches to managing safety and protecting the health of workers, as well as building an occupational safety system in enterprises. Since 1989, when the European Union adopted the Framework Directive "On the introduction of measures that co-operate with improvements in the safety and health of workers at work" risk assessment has become the cornerstone of the European approach to occupational safety and health. In the 2011, the Federal Law of the Russian Federation introduced in Article 209 of the Labour Code of the Russian Federation amendments and additions relating to the definition of the concepts "professional risk" and "professional risk management", as well as the national standard GOST 12.0.010-2009 "SSBT. Labour protection management systems. Hazard identification and risk assessment" [4].

3. Results and discussion

The content of the new approach to ensuring occupational safety on the basis of occupational risk management is that the subject of practical actions should be both the events of damage to health and potential, not yet manifested occupational hazards, which are implicitly hidden and must be recognized beforehand, evaluate and minimize or eliminate. Practice shows that occupational hazards and hazards, which are not explicit, constitute a more significant array of threats than the incidents that have manifested themselves, and their recognition, assessment and elimination have a more significant preventive effect than responding to already occurring facts of damage to health. In addition, direct control of the parameters of the production process, which is carried out in its course in continuous mode, is more effective than monitoring the results of the same process, performed only at the last stage of the event.

One of the important methodological features of the modern model of labour protection is a change in ideas about the nature of the origin of the production incident. In the framework of traditional approaches, the causes of damage to the health of employees are the most often recognized subjective reasons due to their violation of labour protection requirements. At the worker-executive level, such a version has the right to exist, but for the production manager, it does not provide an opportunity to effectively carry out preventive measures, assess, record and manage professional risks.

The ideology of today's production safety and health management system is that the main causes of accidents and occupational diseases are objective reasons, namely, the presence of a certain level of occupational hazards and hazards in each workplace. When hazards and hazards appear, they cause damage to a worker of varying degrees of severity, including loss of ability to work. The more hazards and hazards in production, the more accidents, professional and work-related diseases, as a rule, occur there [5].

Dangerous and harmful production factors can be in two different states - explicitly or hidden or implicitly. The obvious type of finding includes those dangers and hazards that manifested themselves, caused appropriate damage to human health, and in the process of investigating the incident, negative factors were documented and thus legalized. The degree of damage to the employee - light, moderate or severe - should not be taken into account in the procedure for registering a particular hazard or hazard. What is important is the fact of the manifestation of production hazards and hazards, and not the magnitude of the negative effect of their impact.

Those occupational hazards and hazards that are present in the labour process, but have not yet manifested themselves, externally did not affect the human body and, accordingly, did not cause its damage, are classified as dangerous and harmful factors that are in a hidden, implicit state. They pose a particular threat to labour safety due to the fact that, firstly, they are much greater than the dangers of the already manifested and harmful to the health of the employee, secondly, they are in an invisible, non-obvious state and they must be somehow difficult ways to identify and, thirdly, their potential impact on personnel may have more unexpected data and the most unpredictable consequences [5].

The implementation of a professional risk management methodology radically changes the approaches to assessing the state of the workplace. In the traditional model of labour protection, the working conditions of a person were determined by comparing the parameters of the production environment with the permissible indicators established by regulatory documents. This is the so-called centralized regulatory regulation of working conditions, based on the requirements of state regulations. The existing procedure gave the federal authorities the monopoly right to establish what should be considered a dangerous phenomenon or process and what should not be considered dangerous. In the event that the workplace was declared dangerous or harmful, the state established benefits and compensation for workers in harmful working conditions.

The introduction of a professional risk management system means a transition from regulatory state regulation from above to assessing the magnitude of risk in those places and areas where they arise and exist, i.e. directly in the production team. Today, in practice, the concept of “acceptable risk” is used, i.e. risk, the level of which is recognized as acceptable by state institutions, as well as by the production community, public organizations and directly working teams. The new order assumes that most of the powers and responsibilities for the state of working conditions from state structures will be transferred to employers and subjects of the Russian Federation. The establishment of risk acceptable, acceptable or unacceptable is not carried out by the regulatory regulation from above, but by the employer on the basis of the technical and economic possibilities of the enterprise. Decentralization of management in the field of industrial safety increases the responsibility of managers for occupational safety, reduces the prerequisites for the formal implementation of state labour protection requirements, it largely reflects the market economy and corresponds to the nature of entrepreneurial activity as the risky activity [6]. Three important points are characteristic features of a professional risk management system. First, management activities are a cyclical sequence of performing interrelated practical actions, known as the "Deming-Shewhart cycle" or as the PDCA cycle (figure 1). The content of the cycle includes a set of procedures, divided into several stages: Planning (plan) - Implementation (do) - Examining the result (check) - Action (act). The cycle of measures to improve working conditions has a circular character, so the last stage is not the final action of the cycle, after its completion new tasks are set to reduce the level of occupational hazards. Thus, the PDCA cycle is aimed at a constant desire to improve working conditions, to consistently systematic work to reduce the value of occupational risk.

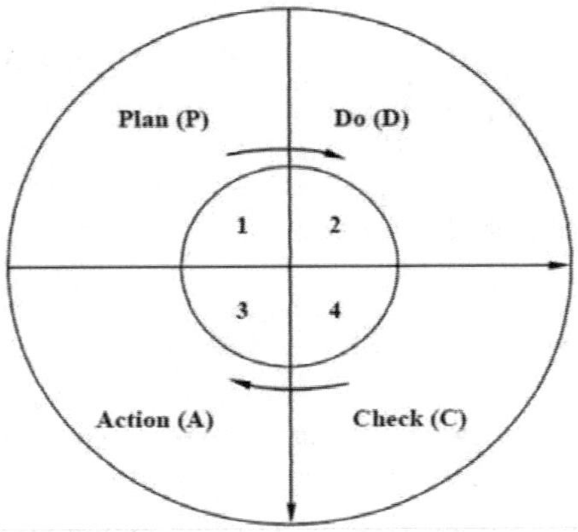

Figure 1. Scheme of the Deming-Shewhart cycle.

Secondly, the consistent implementation of identical cyclical procedures implements in solving management problems the so-called "process approach", which should be carried out continuously, thereby creating conditions for the continuous improvement of the labour protection management system. And, thirdly, upon completion of the work of the last stage, a new cycle of the same practical actions should be started at the same workplaces, but with new goals and objectives for reducing the risks that emerged from the analysis of the last stage.

In general, the PDCA methodology is a kind of algorithm for the actions of the management team to achieve the planned goals in the field of labour protection. The process includes the following steps:

1. Planning (plan). In accordance with the organization's policy in the field of labour protection, the objectives of future actions are established, and a list of necessary measures is developed for them with an indication of the timeframe and responsible officials. The objectives of the work are identified according to the results of the analysis of the actual state of working conditions, carried out on the basis of the obtained data on the identification and assessment of occupational risks.

2. Implementation (do). The plan of measures to achieve the objectives of the policy in the field of labour protection contains deadlines and responsible for implementation. A feature of the risk management methodology is that the responsibility for the implementation of risk mitigation measures must be handled by the leading person who creates these risks. For the employee, the risks are created by the employer and the immediate supervisor of the production site, i.e. the person who provides him with a workplace, an instrument, ensures adequate sanitary and hygienic conditions, etc. Linear management personnel using the labour protection service, which does not full, but coordinates and

IOP Conf. Series: Materials Science and Engineering **728** (2020) 012015 doi:10.1088/1757-899X/728/1/012015

consults the other services, implements measures in the order in which the risks were ranked by importance [7].

3. Examination of the result (check). Stage of the cycle at which the implementation of the work of the 2nd stage is monitored. It measures processes to determine their relevance to the objectives of the program and the organization's labour protection policy. An assessment of the effectiveness of the measures taken is given, the results of the implementation are analysed, the shortcomings of the work program or the shortcomings of its implementation are identified. Control actions can initiate adjustments to both the work program and changes in the formulation of the objectives of management procedures. Promising is the development of internal standards of the organization, setting the so-called "technically achievable standards", more secure than state standards.

4. Action (act). In the Deming-Shewhart cycle, this stage refers to the activities of the company's top management, to the implementation of management functions, in contrast to the first three stages, characterized as the execution of planned activities. The employer, having the highest authority in the enterprise, on the basis of a comprehensive analysis should evaluate the first cycle of the process approach and, if necessary, make adjustments to work plans, declared goals to improve working conditions, as well as formulate new tasks and set new goals to reduce risks to be achieved when implementing actions in the next PDCA cycle.

The employer is a key element in a new management system in an enterprise, not only because it performs the functions of the first head, but also because the current legislation of the Russian Federation has overall responsibility for organizing work to ensure the required level of labour protection on it. In the traditional labour protection management system, these duties were eroded and not sufficiently personified, the labour protection engineer and the corresponding service of the enterprise carried out the main work on the prevention of occupational injuries and occupational diseases, and was supervised by the chief engineer.

In management theory, the principle of «Communicating vessel » is used, which suggests that the effectiveness of management decisions depends on how much the level of responsibility of the employee corresponds to the level of his capabilities or authority (Figure 2). The employer has the highest authority in the organization, he manages financial, material, human and other resources, so he has the opportunity within the existing authority to most successfully solve the most complex tasks, including ensuring comfort and safety of workplaces. Bearing personal responsibility for the general state of safety in the enterprise, the employer delegates some of its powers to the lower management levels, whose managers are responsible for working conditions at the production sites under their jurisdiction. Thus, the first head organizes the work on the management of labour protection in the organization and, secondly, ensures the safe work of his subordinates by creating appropriate working conditions.

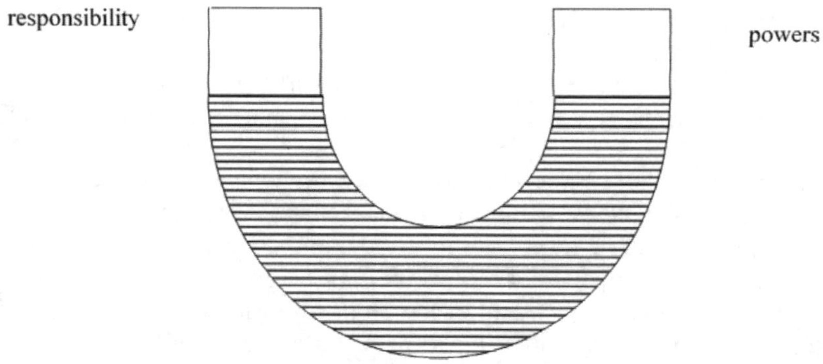

Figure 2. The «Communicating vessel» in the management theory.

The chief assistant of the employer in the implementation of their tasks is an occupational safety specialist. In accordance with the nature of the activities, it is not part of the technological management system of the enterprise and does not have authority over any production site. In contrast to the first manager, the labour protection specialist has limited powers, he is not able to give orders to managerial personnel or manage financial resources in an orderly manner. The labour protection specialist performs essentially the functions of a headquarters employee - he as an assistant and responsible consultant in the management recommends, informs, organizes and coordinates work on occupational safety. This status does not reduce the authority of the labour protection specialist; on the contrary, it places increased demands on his competence, professionalism, and human qualities [7-10].

4. Conclusions

Industrial injuries, occupational and occupational diseases are a serious socio-economic problem for each state. The damage to health that the production process receives is able-bodied and, therefore, the most active part of the population, negatively affects the preservation of the country's human resources, the decrease in life expectancy, and the worsening demographic situation. At the same time, all four sides of the accident or illness that occurred - the injured worker, the employer, the society and the state - incur significant direct and indirect economic losses.

Labour protection does not begin at the machine, not in the workplace. To a much greater extent, occupational safety, like all aspects of production, is laid down in decisions of officials who have a corresponding worldview and implement the modern system of occupational safety and health management. It is expressed in identifying and recognizing dangerous and harmful production factors, analysing and assessing occupational risks, developing organizational and technical solutions for minimizing the identified risks, and in implementing the planned measures with wide participation of the entire workforce. The new model of labour protection management should be an effective mechanism to ensure safe and comfortable working conditions, preserving the employee's proper working capacity throughout his work experience [11-15].

The task that has to be most often solved in the process of designing the organization of repair and construction production is the formation and calculation of irregular streams with unequal sizes of private work fronts. This is due to the heterogeneity of residential buildings undergoing repairs, as well as different volumes of work on the same grips, due to different technical conditions of the same structures, elements, systems in different parts of the buildings being repaired.
The spatial parameters of the stream during the overhaul of buildings almost always determine its irregularity [16-20].

This analysis allows us to argue that in the repair and construction industry the main method of organizing work will be a specialized stream, the products of which are the same type of structural elements, or the same types of work. At the same time, the more buildings are simultaneously under repair, the more efficient and long-term private streams will be, which once again confirms the advisability of overhauling multi-apartment residential buildings by the group method.

References

[1] T. A. Golikova, "On measures aimed at improving working conditions, preserving life and health of employees. Report on the meeting of the Government of the Russian Federation on October 27, 2011," *Occupational safety and health in construction*, no. 1, pp. 7-11, 2012.
[2] GOST 12.0.230 – 2007, *SSBT. Occupational safety management systems. General regulations*, Moscow, Standartinform, 2007.
[3] A. G. Fedorets, "Management of industrial safety and risk assessment," Collection of articles - "IBT", 2012.

IOP Conf. Series: Materials Science and Engineering **728** (2020) 012015 doi:10.1088/1757-899X/728/1/012015

[4] GOST R 12.0.010-2009, *SSBT. Occupational safety management systems. Definition hazards and risk assessment.* Moscow, Standartinform, 2011.
[5] E. B. Sugak, "The nature of occupational injuries in the aspect of managing pro-fess-regional risks," *Life safety*, no. 7, pp. 3-7, 2015.
[6] A.V. Moskvichev, N.I. Simonova, S. V. Vikhrov, and V. V. Ivanov, "The system of labour protection based on the concept of occupational risk management," *Occupational safety and safety in construction*, no. 5, pp. 56-60, 2016.
[7] E. B. Sugak, "On some of the official duties of security specialists Labour in Germany," *Man and work*, no. 10, pp. 48-51, 2013.
[8] GOST R 54934 – 2012, *Occupational safety and health management systems*, Moscow, Standartinform, 2012.
[9] A. Fedorets, *Methodical bases of quantitative estimation of production - Energy security in documents and facts*, no. **2**, pp. 10-16, 2008.
[10] E. B. Sugak, "On the modern model of labour protection," *Sb. Theses of Int. scientific conference "Integration, partnership and innovations in building science and education*," Moscow, MSUCE, pp.185-186, 2013.
[11] E. B. Sugak, "Methodological and organizational problems of professional risk management in construction," *MATEC Web of Conferences*, vol. 196, 04056, 2018.
[12] E. B. Sugak, "Occupational Risks Management as a Basis of Industrial Injuries and Occupational Disease Prevention," *IOP Conference Series: Materials Science and Engineering*, vol. 365, 062038, 2018.
[13] R. A. Litvinov, "Formation of OSH management systems. Tendencies at the national level," *Occupational health and safety at construction*, no. 4, pp. 12-18, 2013.
[14] G. I. Tikhonova, A. N. Churanova, and T. Yu. Gorchakova, "Occupational risk in terms of industrial injuries in Russia," *Occupational medicine and industrial ecology*, no. 4, pp. 9-14, 2012.
[15] E. G. Sugak, "Social Partnership," *Life Safety*, no. 6, pp. 6-10, 2016.
[16] H. W. Heinrich, *Industrial accident prevention; a scientific approach*, 1959.
[17] *The Labor Code of the Russian Federation*, as amended on December 30, 2015.
[18] *A manual for monitoring working conditions in the workplace in industry*, Elmery - Helsinki System - Finnish Institute of Occupational Health, Occupational Safety and Health Administration under the Ministry of Social Affairs and Health of Finland, 2000.
[19] M. Murtonen, "Risk assessment in the workplace - A practical guide," *ILO Subregional Office for Eastern Europe and Central Asia*, 2007.
[20] Guidance on OSH management systems. (ILO-OSH 2001) / (ILO-OSH 2001). Geneva, International Labor Office, 2003.

Disorders of the Green Roof of the Pool Lucenice

Martin Dědič [1], Aleš Kaňkovský [1]

[1] Institute of Technology and Business in České Budějovice, Department of Civil Engineering, Okružní 517/10, 370 01 České Budějovice, Czech Republic

dedic@mail.vstecb.cz

Abstract. The main objective of the article is to inform about the construction-technical survey, which focuses on the existing roofing of a pool area. The pool area is a separate object annexed to the historic farmhouse dating from the mid-19th century. Roofing of the pool area is made of a green roof. The main research issues stem from the building-technical survey focused on the green roof disorders. After detecting the faults, we look for their causes and find out which faults are fatal to the building and which can be easily removed. From the evaluated constructions, we focus primarily on the solution of the green roof in terms of correctly executed roof slope, plating, thermal technical properties of the materials, inherent weight of the structure and execution of individual layers of the composition of the green roof. Furthermore, we focus on determining the compliance of project documentation with the construction. We assess the impact of failures on the structure and operation of the building. In the conclusion we propose measures based on the found failures and possibilities for the solved structure.

1. Introduction

Our task was to find out the condition and defects of the roof cladding and the whole area of the swimming pool building, which is located in the northern part of the eastern wing of the whole farm. Especially in areas where new steel roof structures were made and where these new structures meet the existing structure of the farm. The main objective is to focus on the extent of defects and possible damage, possible emergency and its reasons. We have also considered the possibility of eliminating these disorders.

The survey did not include any other defects in the original farm in Lucenice. This building as a whole is undergoing a gradual reconstruction according to the phasing of individual wings and at the present time it seems to be substantially completed and reconstructed. The original object of the farm dates back to the middle of the 19th century (about 150 years old) and its original, historical structures correspond to the time of its origin, both technically and morally, with the fact that in recent years they have undergone demanding construction work and got new features. The assessed part of the building was realized on the basis of project documentation [1].

2. Construction-technical survey

The basic inspection was carried out in February 2017 and local part of the roof cladding (green roof layer) was also discovered. Therefore, we also checked the external condition of the building and other contexts that relate to the roof cladding in its entirety, not only above the problematic leaks into the building.

It was found out that the whole pool area was realized as a combination of masonry constructions, concrete constructions and these were subsequently supplied with a steel roofing with a proposed layer of composition. The project documentation does not show the scope of supply, but after the owner's interpretation of the building is a system of subcontracting, which delivered the individual building parts. Furthermore, it was found that the entire floor area of the interior of the swimming pool of approx. 6x17m shows a leak into the building in the form of dripping of rusted water. Drip most often occurs at the point of contact of steel rolled beams and galvanized trapezoidal sheet.

At the time of the local investigation, the pool was filled with clean water at a temperature of about 10-12°C and the temperature inside the room was about 15.5°C with a relative humidity of 37-40%. Furthermore, the pool level was covered with a bubble floating pool foil and suspended water buckets were installed at the dripping points. The exterior shell of the pool was finished with a wooden façade and a green extensive roof with some backlog in the form of missing plumbing elements at the attic, and in the absence of foil waterproofing, rainwater drainage, etc. Inside the pool area steel pool roofing is designed only as a technical appearance with exposed steel structures and trapezoidal galvanized sheet.

When examining the lower face of the trapezoidal sheets of the roof cladding, it can be stated that due to the pool environment and the solution of the space as such, there is clearly visible excessive condensation of water vapor, subsequent precipitation on the surface and gravity flow down a certain slope back into the pool area. This process is constantly manifested as dripping rusted water from various stabilized places where plastic buckets are suspended [2].

3. Assessment of research questions

In order to assess the condition of the roof cladding, a local investigation had to be undertaken in order to reach an unequivocal conclusion on the source of the problem by analysing the state of the environment [3].

3.1. Is the roof structure of the pool constructed in accordance with the provided project documentation? What differences, if any?

The submitted project documentation is kept as production in relation to the roof cladding in all its three execution dates. It is therefore not a classic building documentation with all connections and sorting according to the Decree as of the date of drawing up the project documentation [4].

On the basis of the above it can be stated that the roof construction and roof cladding made of steel construction and trapezoidal galvanized sheets are made in accordance with the submitted project documentation. This documentation is created mainly as a production and as such it does not deal with other building contexts and necessary interconnection with the original building and other professions on the construction site [5].

As the only differences, it shows at the prescribed slope of the glazed part of the roof, where the project documentation indicates a slope of 3% and an expert on the spot measured 2.5% slope, which is a negligible difference. Furthermore, there is a difference in the implementation of the green roof stack according to the project as a 100mm substrate, when this stack actually gets more to the edge of the attic glazed roof and also below the neck of the installed ventilation turbines. Furthermore, the axonometry of the pool area according to the submitted project documentation does not solve one actually realized narrow window opening from the hygienic and technological room of the pool

3.2. Do the design and the actual construction of the roof structure take into account the fact that the roof structure is built above the pool? Does the construction of the roof structure allow for proper use of the pool in the long term?

The main problem of project documentation is in the absence of interconnection with classical building documentation. It is this documentation that should give clear answers and set the entry conditions for the solution of the premises. In the pool area under consideration, the main source of the problem is the increased humidity and the subsequent precipitation of water on the face of the roof structure itself or partly in the layers of the designed shell structure. It can be assumed that in the pool area there will be, as now the pool is made of increased relative humidity and further possible aggressive environment in the form of chlorinated water, etc.

These internal influences have caused the simultaneous unexpected behavior of the roof cladding composition. In the construction, water vapor condensation, which is manifested to the surface by dripping of rusted water at the points of contact of trapezoidal sheets, is clearly taking place.

The design and subsequent construction of the roof structure and roof cladding does not take into account the constructional fact of the indoor pool interior, where the class of such an environment referred to for example to protect concrete structures against corrosion is like XD2 - wet and rarely dry (eg swimming pools). The design and actual constructional design could be used in this way if certain input conditions are met, but there are no important measures to avoid environmental impact.

In the long term, such a construction of the roof structure and roof cladding composition is unsuitable for the purposes of the pool area and the current situation without any necessary modifications of the whole situation will continue to deepen. If the investor is satisfied with the buckets installed, this internal corrosion will more attack the galvanized sheet of trapezoidal sheets and outside the composition, where it will degrade the sheet. Of course, the slow corrosion of other supporting steel structures, which are now sufficiently coated, will continue, and in the form of aesthetic maps, the interior will continue to be exposed to dripping water. The current state of the roof cladding and the roof structure therefore prevent the roof clad from being used properly as a defect-free item. The removal of the current situation and the remediation of some of the above-mentioned facts will be highly desirable from the point of view of future operation in the pool area, while maintaining the reality of the swimming pool.

3.3. If the design of the roof structure does not allow for proper use of the pool, are there any structural adjustments that could be made to the existing roofing to enable proper use of the pool? If such construction work exists, what is it?

It is necessary to immediately remove the main causes of water condensation above the trapezoidal metal sheet and, if necessary, to add new constructions and materials composition in parameters according to the original specification, that it is a swimming pool area with use all year round except winter. A more demanding solution would result from the need to replace the entire roof cladding composition, leaving only the supporting steel structure. In such a case, after removing or inspecting all long-term trapezoidal sheets exposed to moisture, it would be advisable to address the new roof cladding as a complex construction activity and to revise all related structures affected by repairs.

The current state of construction of the roof structure and the composition of the roof cladding does not allow for comfortable and full use of the pool area for its original purpose. The situation can only be solved by recommended completion of the composition towards the interior of the pool and revision of existing compositions in general. The desired goal of using the pool area for its main purpose is therefore not necessary to achieve by replacing all elements of the roof structure, but it is only possible to revise the composition of the roof cladding. It is definitely necessary to follow the recommendations resulting from the architectural analysis of the situation and the subsequent proposed building modifications.

First, it is necessary to ensure sufficient forced dehumidification of the entire pool area by an oversized and correctly positioned pool dehumidifier or condensation dehumidifier with recuperation. The designed and realized spontaneous ventilation turbines cannot function in the pool area and can be a sufficient substitute for forced ventilation and dehumidification.

Another necessary measure is to partially cover the water surface with bubble foil or a possible swimming pool shutter when the pool is not in use. By preventing evaporation from the water surface, the effect of temperature differences between the pool water and the pool environment is eliminated.

Furthermore, it is ideal to work with the water temperature in the pool depending on the temperature of the pool area. It would therefore be advisable to consider a comprehensive heating control and regulation depending on the set temperature of the pool water and the current state of the indoor and outdoor environment.

Last but not least, it would be good to revise the current state of the composition, primarily from the interior of the pool area at the site of today's visible ceiling in the form of trapezoidal sheet metal. It is the critical point of the composition that is located above the trapezoidal sheet and vapor-proof foil. When designing the composition above the pool area defined by the final parameters, the overall thermal insulation layer should achieve such insulation that the thermal conductivity coefficient of the structure U does not exceed 0.24 $W/(m^2.K)$. A good vapor barrier must be created under the thermal insulator so that the condensation in a 1 m^2 insulated construction does not exceed 0.1 $kg/m^2/year$. It must also be ensured that the condensation formed is not greater than the possibility of evaporation. Also, the composition as a whole should be airtight. The present state of all or substantial condensation takes place between the trapezoidal sheet layer and the vapor barrier, and in the form of running water, this is reflected in the overlapping sections of the individual trapezoidal sheet metal sections. Such a poorly chosen vapor barrier film and its placement complicates the basic consideration in the climate in the pool area (eg + 30°C/75% humidity), then the dew point of the structure is just behind the vapor barrier (in our case trapezoidal sheet). The same climate should be ensured in the vapor barrier behind the ceiling layer as in the indoor environment. Thus, the ceiling layer allows the air heated from the interior to flow beyond the ceiling layer into the gap between the vapor barrier and the ceiling layer. Thus, in contrast to the present state, a closed air gap would not form between the vapor barrier creating an additional thermal insulation layer of the structure.

3.4. Is the cause of leakage into the building the so-called "green roof" and its implementation? Or to what extent?

According to the project documentation, the "green roof" is defined by the composition as the last layer of the composition in the definition of "green roof stack (100mm substrate)". These tracks are described in sections and are not specified anywhere else. From this information we conclude that there are many types of "green roofs" and it can only be said from the local survey that this is the most economical variant of a green roof with a single layer with an extensive substrate.

It is apparent from the above answers that the main problem of the roof is not leaking into the building, but an inappropriate design of the roof cladding. It can therefore be said that the project documentation does not allow to know very much what the green roof is and therefore no related details in certain complicated places are further addressed. We only state that the implementation of about 4 layers of "green roof" is not only in 100 mm substrate, but the composition contains more elements, which are finally implemented as higher. Compared to the project documentation, the project implemented a layer for the "green roof" as a levelled layer in the attic and the flat glazing collar. As a result, the ventilation turbines practically flush. In the case of ventilation turbines, it happens that the turbine stops in the event of snow and when it grows to the desired size of planted stalks-herb-grass, so the turbine finally stops and stops venting. The fact is that the ventilation design is completely inadequate and even designed in such a wrong way, because it rather causes permanent unwanted roof perforation and creates a local thermal bridge, and at this point there is some other induced condensation and freezing instead of ventilation and properly designed.

4. Conclusions

Given the available information and their evaluation, we are able to answer the questions as follows.

The roof structure of the pool is made in accordance with the project documentation provided. This documentation is created mainly as a production and as such it does not deal with other building contexts and necessary interconnection with the original building and other professions on the construction site. Implementation compared to the project documentation shows negligible differences that do not significantly affect the main and material errors.

The design and subsequent realization of the roof structure and roof cladding does not take into account the fact that the roof structure and roof cladding are built above the pool. The roof construction and roof cladding do not allow proper use of the pool in the long run.

The roof structure allows the building to be used as a construction base for the pool area. Furthermore, the roof cladding and its composition do not allow the use of the pool area for its main purpose without the need for the constructional modifications outlined in this assessment point. Thus, there are building modifications that could be made to the existing roof structure to allow proper use of the pool.

The green roof is not the main cause of leakage. Only the construction of the roof in terms of general technical requirements for construction is questionable in places where the details can be solved completely differently and better.

To ensure the perfect use of the swimming pool area in the building, the following modifications must be made:

- Based on the current state of the roof structure and roof cladding, it would be good to stabilize the state of the indoor environment (ventilation, dehumidification, heating, etc.)

- Ideally, re-design the project implementation documentation for state-of-the-art repairs to identify corrective actions and identify clear points of solution. There will probably be more significant system solutions for the new supplier of remediation work. The new project documentation will therefore contain consistent comprehensive solutions before the actual step of remediation of the current building modifications of the swimming pool area.

References

[1] M. Dědič, "Analysis of Historical Residence on Terms of the Current State," *IOP Conference Series: Material Science and Engineering*, vol. 471, issue 8, 082033, 2019.
[2] J. Kaiser, "Process modeling for BIM," *CESB 2016 - Central Europe Towards Sustainable Building 2016: Innovations for Sustainable Future*, pp 781-788, 2016
[3] K. Prušková, M. Dědič, J. Kaiser, "Possibilities of Using Modern Technologies and Creation of the Current Project Documentation Leading to the Optimal Management of the Building for Sustainable Development," *IOP Conference Series: Earth and Environmental Science*, vol. 290, issue 1, 012058, 2019.
[4] M. Dědič, "Defects of roofing and roof structure – Apartment house Kaplice," *IOP Conference*

Series: Material Science and Engineering, vol. 603, issue 5, 052088, 2019.

[5] J. Myslin, J. Kaiser, "Priority of Requirement and Method for Definition of Project Scope," *Proceedings of the 29th International Business Information Management Association Conference - Education Excellence and Innovation Management through Vision 2020: From Regional Development Sustainability to Global Economic Growth*, pp. 4002-4013, 2017.

IOP Conf. Series: Materials Science and Engineering **728** (2020) 012017 doi:10.1088/1757-899X/728/1/012017

Experimental Measurements of Wooden Solid Log Wall

Petr Kacálek [1], Tomáš Petříček [1]

[1] Brno University of Technology, Faculty of Civil Engineering, Institute of Building Structures, Veveří 95, 602 00 Brno, Czech Republic

kacalek.p@fce.vutbr.cz

Abstract. The contribution brings findings accrued from alternative design of log-wall's lateral joint. The lateral joints present without question a thermal bridge highly repeating in the construction of external log wall. Field of the research was a theoretical design of wall models, simulation in software ANSYS and comparison theoretical results with results gained from the laboratory measurements of wall samples. Bed joints of wooden solid log walls are still a problematic detail in terms of thermal technology. By reason of weakening in the place of lateral joint and the value of coefficient of thermal conductivity, the solid wooden log wall does not meet requirements of Czech Standards ČSN 73 0540-2 posed to the value of wall's thermal resistance. That is why we have focused on measurement of surface temperature, which is also one of the classified factors in term of construction quality. The project particularly focused on monitoring of surface temperatures in the areas of critical details – bed joints.

1. Introduction

The paper concerns a Canadian style solid timber log house wall in terms of thermal technology, wherein the main goal was to compare construction variants of such wall. The comparison concerned a circumferential wall of a Canadian style log house, using different types of heat insulation. The wall models were simulated in ANSYS software. Due to weak spots near bed joints and the wooden thermal conductivity coefficient, the solid wooden wall has difficulty complying the existing strict requirements of the standard ČSN 73 0540-2 [1] for heat transmission coefficient value U. Therefore, surface temperature and the internal surface temperature factor were concerned this time. They are one of the quality evaluation factors for structures and the internal environment in buildings.

2. Bed joints

Bed joints are regarded as a weak spot especially in peripheral solid timber based walls, where they are repeated all over the surface of the vertical structure. Such details of bed joints are however more problematic in case of log wood wall in comparison to those made from lumber, since the weakening of bed joints is much more seen.

The connection of logs must be perfect in all aspects. It must create a permanent barrier between the interior and exterior of the building. Although the logs for log houses are carefully selected (the logs can show any torque in fibers) after drying it may come to a deformation, thus a change in their shape. Such changes do cause a leakage and weakening of the joints. Materials for the final sealing of all bed joints must have certain insulating properties, both in terms of thermal resistance and also in

IOP Conf. Series: Materials Science and Engineering **728** (2020) 012017 doi:10.1088/1757-899X/728/1/012017

terms of air infiltration. Nevertheless the material must also exhibit shape memory and must be able to deal with creep.

Earlier the bed joints between the logs were stuffed out by moss which was then covered by a layer of clay screed. The surface of the clay screed was then dyed white or coated by a different material. The system in overall acted as a sealant. As the years moved on and it came to a development of newer materials and technologies polyurethane foam became the number 1 sealant used. Nonetheless, polyurethane foam when hardened is dimensionally stable and is unable to respond to the volumetric changes of wood. Consequently cracks are formed which must be additionally sealed to prevent heat loss. Concurrently it was found out that the best solution would be to and also is to use a two-stage isolation, where the thermally insulating capabilities are maintained by mineral or sheep wool, and the air leakage is prevented by memory tapes based between the logs on the external and internal side of the wall likewise.

Mineral wool is positioned at contiguous locations of the longitudinal groove placed between logs used to make a structure separating a heated environment from that of a non-heated one, since the grooves may have a different shape and size. The memory tapes are then made from permanently elastic materials based on specially treated foam impregnated with a substance that increases its resistance to UV radiation and reduces water absorption. Memory tapes do also secure the structures against seeping of water, air infiltration and also against the entry of insects into the longitudinal groove.

3. Internal wall surface temperature factor evaluation

The calculation of the internal wall surface factor fRsi and internal surface temperature (°C) was performed for other possible comparisons. The comparison selected a bed joint of a log house wall, so-called double insulated, which means that a storage tape ILLBRUCK is circumferentially applied on the joint, and the joint area is filled with heat insulation – mineral wool. Total joint width was designed to be 15 cm. Meeting of the requirement $f_{Rs}i \geq f_{Rsi,N}$ was monitored.

The calculations make it clear that the requirement f_{Rsi} (0.936) $\geq f_{Rsi,N}$ (0.808) is met for the critical detail of the circumferential log house wall – bed joint. Internal surface temperature $\Theta_{si} = +\ 17.7$ °C.

4. Assessed structures

We created an experimental sample of timber log wall using 3 m long logs with diameter of 350 mm. Spruce logs were based on asphalt sheets [2,3]. We tested 3 types of different thermal insulation installed in the bed joints. Memory tapes were used in all variants, only the insulating material was changing:

- In 1st variant there was only the memory tape without any added thermal insulation in the joint;
- In 2nd variant the joint was filled with mineral wool;
- In 3rd variant the joint was filled with fleece.

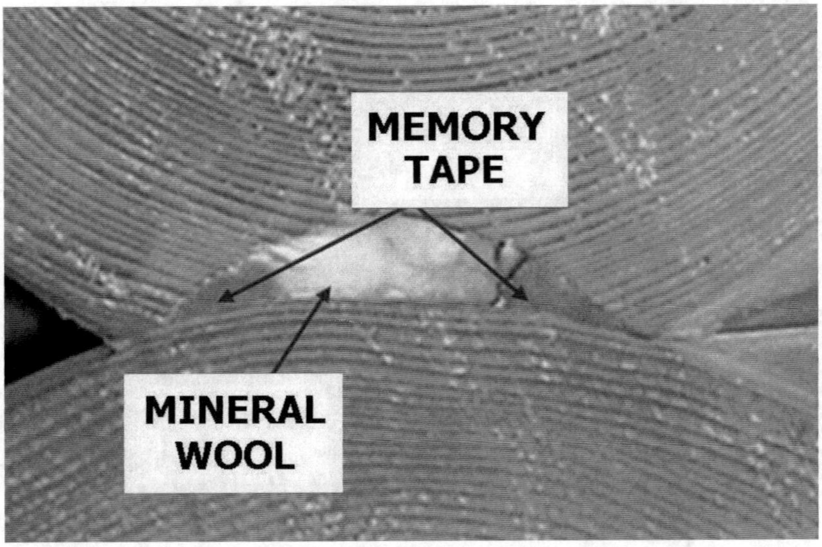

Figure 1. Schema of the assessed log wall's bed joint.

5. Results of the ANSYS assessment

We compared the lowest surface temperatures at the critical location of any solid timber log wall – the bed joints. Installing the thermal insulation into the bed joints has a positive effect on the results of the assessment. The difference between used insulation materials (mineral wool and fleece) is not significant. In 1st variant (without added thermal insulation) the lowest surface temperature $\theta = + 13.735°C$. In 2nd variant (mineral wool) the lowest surface temperature in the bed joint was $\theta = + 13.995°C$. In the 3rd variant (fleece) the ANSYS calculated that the lowest surface temperature was $\theta = + 13.999°C$.

6. Results of the experimental measurements

The experiments were carried out in the laboratory of Institute of Building Structures on the sample of timber log wall. One side of the wall was exposed to inner climate of the laboratory (simulating common indoor environment), on the other side we simulated the outdoor environment (temperatures between - 15 and - 18°C). The record contains a total of 13 000 values for each temperature.

The cooling device (the evaporator from Klimastar s.r.o. Cologne) was set to maximum temperature - 15°C with slight decreases in 30 minute intervals.

The experimental measurement does not take into account the effects of the base, atypical details of objects, etc. [4, 5]

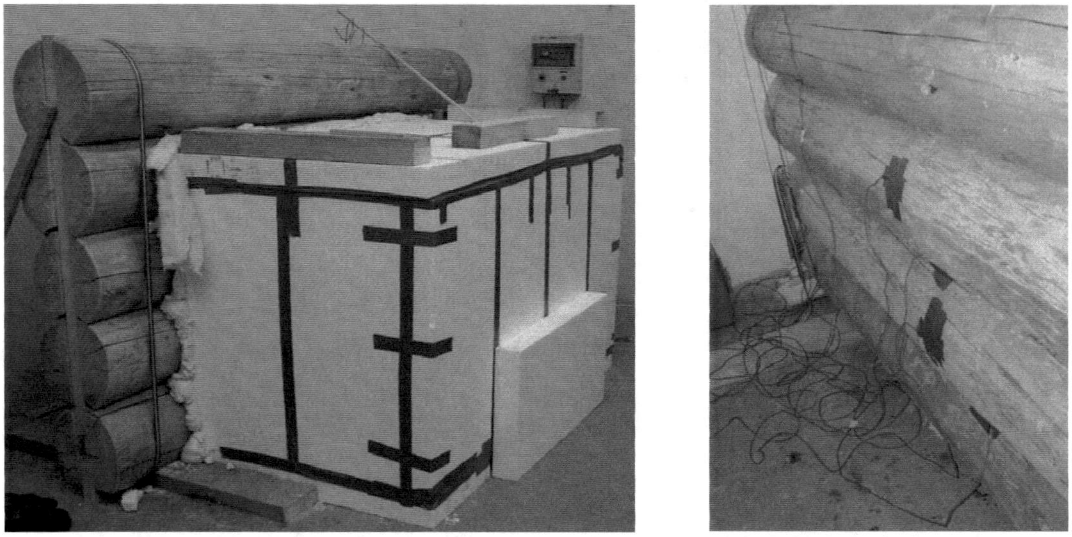

Figure 2. Experimental sample of timber log wall with installed cooling chamber and measuring equipment.

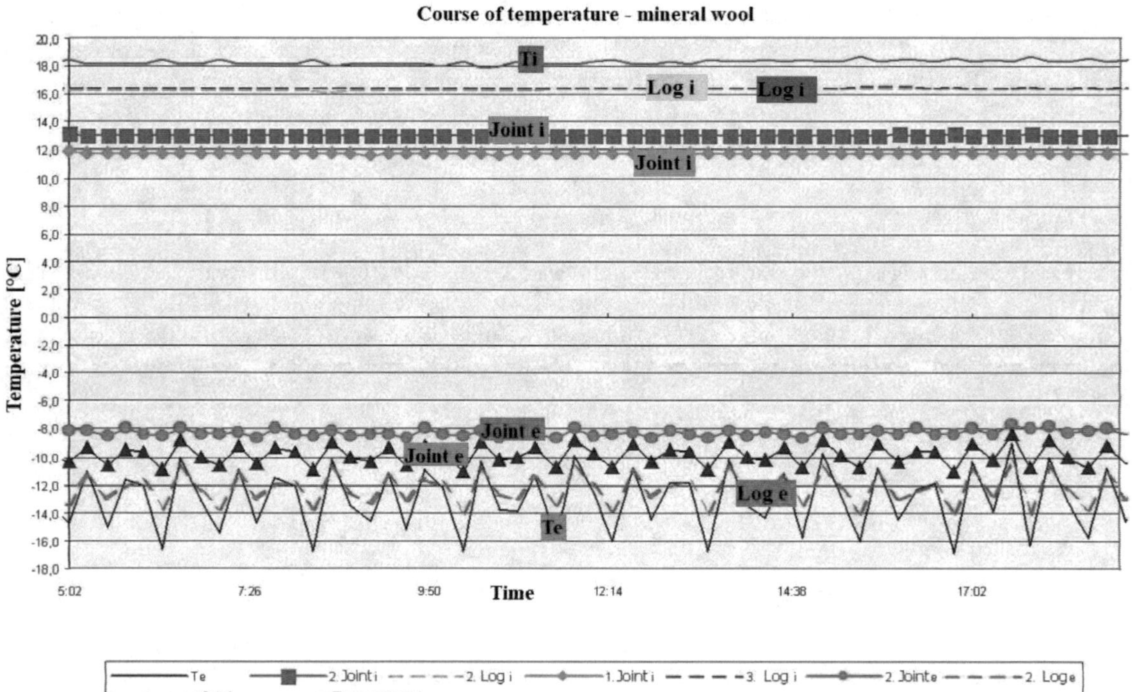

Figure 3. Results of the temperature measurements on both sides of the wall

Figure 3 clearly shows variations in surface temperatures on the "outer" side of the experimental wall, depending on the changes of the surrounding air temperatures. On the "inner" surface of the wall the temperature changed only negligibly.

7. Conclusions

To conclude, solid wooden structures fail to comply with a thermal technological standard ČSN 730540-2 requirement for heat transmission coefficient value U, which may have energy impacts in relation to higher demands to heating of such buildings. Regarding the capacity of wood to sequester CO2, however, a log house from solid wood is a low-energy house in terms of the so-called grey energy spent on construction materials during their production and inclusion in the structure. Solid wooden houses also have a range of positive properties. They are beautiful in appearance and they can create a pleasant atmosphere in the interior thanks to wood specific properties.

Recently the energy efficiency of buildings moves from assessing the energy performance of buildings only according to the thermal-technical point of view to assessing the energy requirements of the building's whole life cycle. This includes the energy consumed during the production of materials, construction process and the use of the building. From this perspective the log buildings are better than common masonry buildings, because wood is available renewable material, the construction of the log building (including proper design of structural details) is simple and the use of such building is inexpensive too.

Deep focus onto the quality of every details execution is required for wooden buildings. Every single cleft between two logs must be filled with sufficient amount of thermal insulation including careful edge memory strips sticking. The correct foundation of the first log is the base, especially it is very important to interrupt the hydro-physical stress and focus onto the proper design of the water proof layer [6].

Acknowledgment(s)

This paper has been produced within the project FAST-S-19-6045 - Study of physical processes in critical details of building envelope structures with near zero energy consumption.

References

[1] CSN 73 0540-2. *Thermal protection of buildings – Part 2: Requirements*, Czech Office for Standards, Metrology and Testing, Prague, 2005, 2007, 2011.
[2] J. Plachý, V. Petránek, and Z. Caha, "Bitumen Substance Analysis," *Advanced Materials Research*, vol. 897, pp. 103-106, 2014.
[3] T. Petříček, P. Kacálek, and T. Hlavačka, "Characteristics of the mechanically fastened joints of one-layer bitumen sheets," *MATEC Web of Conferences*, vol. 93, 03007, 2016.
[4] P. Kacálek, T. Petříček, and R. Smolka, "The Effects of the Soils Thermal Field onto the Energy Performance of Buildings," *Advanced Materials Research*, vol. 899, pp. 58-61, 2014.
[5] R. Smolka, J. Sobotka, and J. Šulistová, "Design proposal for door thresholds," *MATEC Web of Conferences*, vol. 93, 01009, 2016.
[6] J. Sobotka, and R. Smolka, "Use of EMW radiation in the building industry at defects in buildings," *MATEC Web of Conferences*, vol. 93, 01008, 2016.

IOP Conf. Series: Materials Science and Engineering **728** (2020) 012018 doi:10.1088/1757-899X/728/1/012018

Reduction of Light Transmission by Glazing with Atmospheric Pollutants

Petra Machová [1], František Vajkay [1]

[1] University of Technology in Brno, Faculty of Civil Engineering, 602 00, Brno, Czech Republic

petran.3011@gmail.com

Abstract. Daylight has been used for centuries as the main source of light in interiors and has always been an implicit part of architecture ever since buildings exist. Daylight is an irreplaceable environmental factor and is a prerequisite for a healthy stay in buildings. Ensuring daylight access to the interior is a challenge when designing buildings. The aim is to provide enough daylight for the visual work that indoor building users do. The deposition of air pollutants on the glazing can significantly affect the transmission of daylight into buildings located in urban areas. This study shows the impact of air pollution on glazing permeability in three cities in the Czech Republic. The throughput was measured on a total of 9 windows for half a year. Air purity, exposure to rain, the slope of glazing and use of rooms are factors influencing the final permeability of the glazing. Where glazing is exposed to rain, dirt on the outside of the windows tends to accumulate. The final reduction in throughput is usually 4-8%, with some exceptions.

1. Introduction

Daylight in cities is reduced by air pollutants and debris deposited on the surface of buildings. Air pollution causes attenuation of incoming visible radiation. The deposited dirt reduces the reflectance and permeability of building materials.

Glass contamination means visual difficulties caused by the darkening of exposed surfaces by deposition of atmospheric particles [1]. Pollutants are primary particles - fine solid or liquid substances such as smoke particles or dust consisting of alumina, calcium fluoride, cement, coal, grains, limestone, metals, ores, rocks and wood [2]. Pollution has been shown to occur on surfaces exposed to direct rain, indicating that rain washing is not fully effective [3].

Pollution has mostly been studied on glass exposed environments in urban environments. The results of these investigations show that surface deposits consist of mixtures of insoluble (solid residues, fragments rich in metals and black carbon) and soluble (salts rich in S, C and N and organic species of organic matter particles) particles [4-6] and their nature is relatively representative of the abundance of airborne particles present in exposure areas [6].

Between 1999 and 2003, several field surveys were conducted in the UK and Singapore [7-9]. The decrease in daylight transmission caused by the deposition of pollutants in urban buildings was measured here. In the UK, a 4-8% reduction in permeability for commercial buildings in a clean environment was measured, while further research [8] reported diffuse permeability values for vertical windows typically up to 10%. A survey conducted in Singapore that has a high density and tropical humid climate indicates a reduction in overall throughput ranging from 9% to 36% for vertical and horizontal windows [9].

This work deals with air pollution and glazing in the Czech Republic, specifically in three cities and different types of buildings. Glass permeability measurements were performed once a week on 9 windows. These data formed the basis of the project.

2. Air pollution and glass permeability measurement

Airborne dust is small particles of various substances that are so light that it takes a very long time to settle on the surface, and due to this property, the term "airborne dust" known as PM has become known. We distinguish categories PM_{10}, $PM_{2,5}$ a $PM_{1,0}$, by particle size. For example, PM_{10} are particles up to 10 microns = 0.001 mm. The smaller the particle diameter, the longer it stays in the air. PM_{10}-size particles “fly” in the air for several hours, $PM_{1,0}$ and several weeks until they are flushed out by rain.

The airborne dust consists mostly of sulphates, ammonium salts, carbon, some metals, nitrates, or even volatile organic compounds or polyaromatic hydrocarbons [10].

Outside air pollution is low for dust fall below 50 $Mg \cdot km^{-2}$ per year and high for fall above 200 $Mg \cdot km^{-2}$ per year. Indoor air pollution is considered medium in clean workshops, warehouses, etc., large with significant sources of dust in the interior - eg dusty workshops [11].

In the Czech Republic, the 24-hour limit of 50 micrograms per m^3 applies to PM_{10} airborne dust, which limit can be exceeded 35 times a year. Another applicable limit sets the highest average concentration for the whole year at 40 micrograms. However, according to the latest studies, lower concentrations of airborne dust may also have harmful health effects.

Regardless, the limit value for airborne dust is exceeded in one third of the territory of the Czech Republic. However, in this third of the territory, two thirds of the population are exposed to dust and face increased health risks. In the Czech Republic and Poland, the situation is the worst of the European Union. In comparison with air quality in about 30 large European cities in terms of airborne dust pollution, Prague was the worst city ever. The European Union wants to tighten the limit so that the limit of 50 micrograms per cubic meter should not be exceeded more than seven times a year. This has made the existing criterion tolerant of exceeding the set limit thirty-five times more stringent.

Glass permeability measurements were carried out for six months at several locations in the Czech Republic. The first locality was in the center of the regional town of České Budějovice, the second locality was on the outskirts of Hluboká nad Vltavou and the third locality was in the center of the regional capital Brno. The measurements were performed on several different windows with different orientation. The windows were washed before the first measurement and then the glass permeability was measured once a week.

A sample of 9 windows was selected in the study, representing different atmospheric environments in cities and different types of building designs. The buildings were selected in the city center and on the outskirts. The buildings had transparent single or double glazing, different types of heating, vertical and sloping windows, rooms with different functions (from bedrooms to school areas) and windows at different heights above ground level. The core of the sample were the following buildings: a residential unit in the center of České Budějovice (fourth floor), a family house on the outskirts of Hluboká nad Vltavou (second floor) and a university building in the center of Brno (third floor).

For each window, the permeability was first measured on the cleaned glazing from the interior and exterior, and then weekly on progressively soiling windows. Measurements were made using the Pocket

Detective 2.1 Transmission and Reflectance Meter and the Window Energy Profiler - an Energy Transmission Meter for Testing Operable Windows In-Frame.

3. Results

Figure 1 shows the main result of a field study. It shows the loss of throughput in percent in all measured windows. Said loss of permeability is attenuation due to impurities. It does not include the permeability values of actual glazing materials.

Most values are below 10% and only one value (roof window in the university building) exceeds this pollution, all values are less than 15%.

The findings from data analysis by location and use of the room were as follows:

- The reduction in light caused by impurities inside the glazing was usually at least as great as external contamination. In general, the use of the room was an important factor.
- The amount of external debris on the windows was related to the air quality in the immediate surroundings, especially the concentration of particles in the atmosphere.
- External contaminants were significantly affected by precipitation and slope glazing. Windows near the ground were a little dirtier than windows higher on buildings, horizontal windows were dirtier than vertical windows, the glazing exposed to heavy rain was cleaner.

Figure 1. Reduced glazing permeability due to dirt.

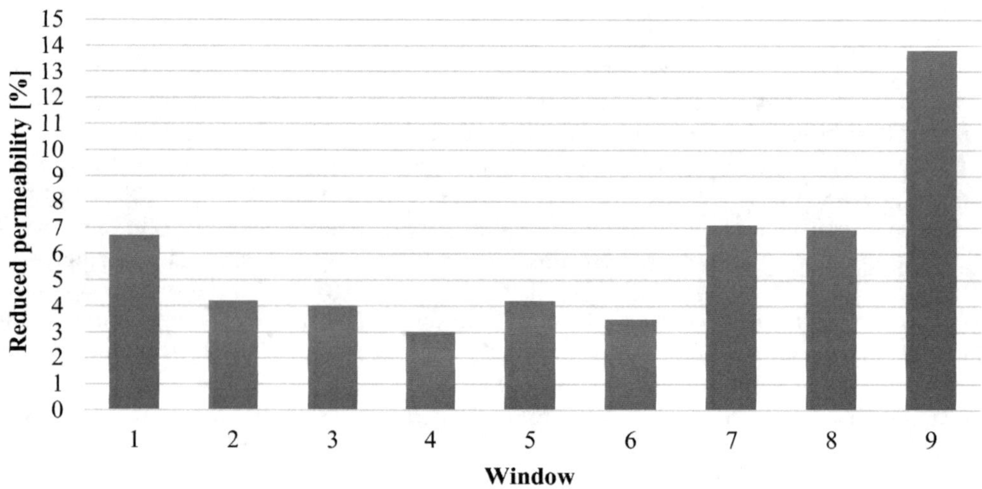

Table 1 and Table 2 are guidelines for the typical loss of daylight transmission due to the build-up of dirt on the glazing. Table 1 shows the percentage loss in relation to room function and building location. Table 2 shows the corrections for the different exposure levels and the type of glazing.

It should be noted that the tables give typical figures based on the measured mean values. The measurement took place in three cities of the Czech Republic. In the future, it is necessary to measure more samples in order to create a table for all room functions and window exposure throughout the Czech Republic.

Table 1. Typical percent reduction in glazing through dirt deposition.

Room function	**Location**	
	Suburban	Urban
Residential Private rooms and common areas, few residents, good maintenance, no smoking	4%	6%
Commercial, educational Rooms used by groups of people, spaces with office equipment	Not measured	7%

Table 2. Corrections to be applied to the values in Table 1 for different exposure conditions.

	Slope of glazing	
Exposure	Vertical	Inclined
Exposed to heavy rain	x 0,5	x 1,5
Normal exposure location	x 1	x 2

The values in the tables can be used as follows:

- Vertical window to the classroom in the city district, normal exposure. From Table 1 we take 7%; multiply x 1 from Table 2, bringing a final reduction of 7%.

4. Conclusion

Daylight is one of the most important environmental factors. The indoor environment of buildings is one of the basic parameters that affect human health and is second only to the lifestyle of the population. Daylight cannot be replaced by artificial sources and together with direct sunlight it is an essential part of our environment as well as air, water, earth's gravity and so on.

Measurement of window transmittance during half a year of non-washing has shown that the pollution of the glazing significantly affects the resulting amount of luminous flux passing into the interior. The greatest difference in permeability after washing and after half a year of non-washing was 7.1% for vertical windows and 13.8% for horizontal windows. The location, orientation, location and material of the glazing have an influence on the resulting amount of pollution for a particular window.

Daylight entering the interior through windows is reduced due to air pollution. The average values of reduction permeability glazing can be found in the tables above. Imposition of impurities is influenced by the function of the room, the concentration of particles in the atmosphere, precipitation and the shape of the building.

Cleaning the exterior windows affects the daylight that permeates through the windows, but has less effect than washing the windows from the interior, because the windows on the exterior side are usually naturally washed by rain. Frequent cleaning of the interior can increase daylight if the interior atmosphere is very humid or contaminated by smoke or dust. Glazing that is shielded from the exterior by overhangs can accumulate more dirt. Here, the relatively low throughput must be compensated for by a larger glass area or by ensuring frequent cleaning.

IOP Conf. Series: Materials Science and Engineering **728** (2020) 012018 doi:10.1088/1757-899X/728/1/012018

Acknowledgment(s)
This paper has been worked out under the project No. FAST-J-19-5964 supported by of Brno University of Technology. The project name is "Vliv znečištění skel na světelnou pohodu v interiérech budov" and identification number is "30791".

References
[1] J. Watt, and R. Hamilton, "The soiling of buildings by air pollution," Air *Pollution Reviews. The Effects of Air Pollution on the Built Environment*, vol. 2, pp. 289-334, 2003.
[2] G. A. Mastekbayeva, and S. Kumar, "Effect of dust on the transmittance of low density polythene glazing in a Tropical climate," *Solar Energy*, vol. 68, pp. 135-141, 2000.
[3] T. Lombardo, A. Chabas, R. A. Lefèvre, and A. Ionescu. „Modelling the soiling of float glass in a polluted atmosphere, " *Glass Technology*, vol. 46, pp. 192-196, 2005.
[4] T. Lombardo, A. Ionescu, R. A. Lefèvre, A. Chabas, P. Ausset, and H. Cachier, "Soiling of silica-soda-lime float glass in urban environment: measurements and modeling," At*mospheric Environment*, vol. 39, pp. 989-997, 2005.
[5] O. Favez, H. Cachier, A. Chabas, P. Ausset, and R. A. Lefèvre, "Crossed optical and chemical evaluation of modern glass soiling in European urban environment," *Atmospheric Environment*, vol. 40, pp. 7192–7204, 2006.
[6] A. Chabas, T. Lombardo, H. Cachier, M. H. Pertuisot, K. Oikonomou, R. Falcone, M. Verità, and F. Geotti-Bianchini, "Behaviour of self-cleaning glass in urban atmosphere," *Building and Environment*, vol. 43, pp.2124-2131, 2008.
[7] P. A. Tregenza, L. Stewart, and S. Sharples, "Reduction of glazing tranmittance by atmospheric pollutants," *Lighting Research and Technology*, vol. 31, pp. 135–138, 1999.
[8] S. Sharples, L. Stewart, P. A. and Tregenza, P. A, "Glazing daylight transmittances: a field survey of windows in urban areas," *Building and Environment*, vol. 36, pp. 503–509, 2001.
[9] M. B. Ullah, J. T. Kurniawan, L. K. Poh, T. K. Wai, and P. A. Tregenza, "Attenuation of diffuse daylight due to dust deposition on glazing in a tropical urban environment," Lighting Research and Technology, vol. 35, pp. 19–29, 2003.
[10] Izolační zasklení. *Polétavý prach* [online]. 2007 [cit. 2019-11-01]. Accesible from: http://hluk.eps.cz/hluk/emise/poletavy-prach-%E2%80%93-neviditelna-hrozba/ (In Czech)-
[11] J. Kanka, "Význam činitelů při výpočtech ve stavební světelné technice*," Světlo*, vol. 3, pp. 40-42, 2008 (In Czech).

AUTHOR INDEX

Cherkina, V M 75
Dedic, Martin 82, 95
Kacálek, Petr 101
Kaganovich, Natalia 48
Kalousek, Lubor 7
Kankovský, Aleš 95
Kraus, Michal 54, 68
Machová, Petra 54, 106
Maltseva, Irina 48
Musilek, Josef 1
Novotný, Marek 42
Novotný, Michal 60
Petrícek, Tomáš 101
Pilný, Ondrej 7
Plachý, Jan 20, 33, 54
Pokrovskaya, E N 14
Senitková, Ingrid Juhásová 68
Shushunova, N S 75
Smolka, Radim 27, 37
Sobotka, Jindrich 27, 37
Sugak, E B 87
Sutliak, Stanislav 20
Sál, Jirí 82
Tkachuk, Konstantin 48
Vajkay, František 106
Vysoká, Jana 33
Všetecka, Jan 60
Žák, Jaroslav 54